Technical Writing Spaces

Writing Spaces: Readings on Writing

Editors
Trace Daniels-Lerberg, University of Utah
Dana Lynn Driscoll, Indiana University of Pennsylvania
Mary K. Stewart, California State University San Marcos
Matthew Vetter, Indiana University of Pennsylvania

Associate Editors
Colin Charlton, University of Texas Rio Grande Valley
Rachel Buck, American University of Sharjah
Xiao Tan, Arizona State University

Copy Editors
Ashley Cerku, Oakland University
Brynn Fitzsimmons, University of Kansas
Analeigh E. Horton, University of Arizona
Jennifer K. Johnson, University of California, Santa Barbara
Heather A. McDonald, American University
Melody Pugh, U.S. Air Force Academy
John Whicker, Fontbonne University

Web Editors
Joshua Daniel-Wariya, Oklahoma State University
Matthew Vetter, Indiana University of Pennsylvania

Social Media Editors
Kath Field-Rothschild, Stanford University

Volumes in *Writing Spaces: Readings on Writing* offer multiple perspectives on a wide range of topics about writing. In each chapter, authors present their unique views, insights, and strategies for writing by addressing the undergraduate reader directly. Drawing on their own experiences, these teachers-as-writers invite students to join in the larger conversation about the craft of writing. Consequently, each essay functions as a standalone text that can easily complement other selected readings in writing or writing-intensive courses across the disciplines at any level.

All volumes in the series are published under a Creative Commons license and available for download at the Writing Spaces website (https://www.writingspaces.org), Parlor Press (https://parlorpress.com/pages/writing-spaces), and the WAC Clearinghouse (http://wac.colostate.edu/).

TECHNICAL WRITING SPACES

Readings on Writing
Volume 6

Edited by

Kirk St. Amant and Pavel Zemliansky

Parlor Press
Anderson, South Carolina
www.parlorpress.com

The WAC Clearinghouse
Fort Collins, Colorado
wac.colostate.edu

Parlor Press LLC, Anderson, South Carolina, USA

© 2024 by Parlor Press and the WAC Clearinghous. Individual essays © 2024 by the respective authors. Unless otherwise stated, these works are licensed under the Creative Commons Attribution-NonCommercial-NoDerivatives 4.0 International License (CC BY-NC-ND 4.0) and are subject to the Writing Spaces Terms of Use. To view a copy of this license, visit http://creativecommons.org/licenses/by-nc-nd/4.0/, email info@creativecommons.org, or send a letter to Creative Commons, PO Box 1866, Mountain View, CA 94042, USA. To view the Writing Spaces Terms of Use, visit http://writingspaces.org/terms-of-use.

All rights reserved.
Printed in the United States of America

S A N: 2 5 4 - 8 8 7 9

Library of Congress Cataloging-in-Publication Data

Writing spaces : readings on writing. Volume 1 / edited by Charles Lowe and Pavel Zemliansky. p. cm. Includes bibliographical references and index.

ISBN 978-1-60235-184-4 (pbk. : alk. paper) -- ISBN 978-1-60235-185-1 (adobe ebook) 1. College readers. 2. English language--Rhetoric. I. Lowe, Charles, 1965- II. Zemliansky, Pavel.

PE1417.W735 2010

808'.0427--dc22

2010019487

1 2 3 4 5

978-1-64317-430-3 (paperback)
978-1-64317-431-0 (pdf)
978-1-64317-432-7 (epub)

Cover design by Colin Charlton.
Printed on acid-free paper.

Parlor Press, LLC is an independent publisher of scholarly and trade titles in print and multimedia formats. This book is available in paper and ebook formats from Parlor Press on the World Wide Web at https://www.parlorpress.com or through online and brick-and-mortar bookstores. It is also available in eBook formats at https://writingspaces.org and https://wac.colostate.edu/. For submission information or to find out about Parlor Press publications, write to Parlor Press, 3015 Brackenberry Drive, Anderson, South Carolina, 29621, or email editor@parlorpress.com.

Contents

1 Who Is the User? Researching Audiences for Technical Documents   *1*
  Emma J. Rose

2 Assessing Sources for Technical Communication Research   *25*
  Therese I. Pennell

3 Last to Be Written, First to Be Read: Writing Memos, Abstracts and Executive Summaries   *45*
  K. Alex Ilyasova

4 Drafting Technical Definitions and Descriptions   *61*
  Quan Zhou

5 Let's Party: Composing a Review of the Literature on a Technical Topic   *87*
  Daniel P. Richards

6 Stronger Together: Collaborative Work in the Technical Writing Classroom   *103*
  Laurence José

7 Worth a Thousand Words: Constructing Visual Arguments in Technical Communication   *123*
  Candice A. Welhausen

8 Technical, Scientific, and Business Presentations: Strategies for Success   *153*
  Darina M. Slattery

9 Writing Technical Content for Online Spaces   *173*
   Yvonne Cleary

10 Social Media as a Space for Today's Technical Communication Work   *193*
   Clinton R. Lanier

11 Introduction to Usability and Usability Testing   *209*
   Felicia Chong and Tammy Rice-Bailey

12 Beyond Audience Analysis: Three Stages of User Experience Research for Technical Writers   *235*
   Joanna Schreiber

13 "Not So Fast": Centering Your Users to Design the Right Solution   *253*
   Candice Lanius and Ryan Weber

14 Basic Approaches to Creating Accessible Documentation Projects: What Is Accessibility, and What Does It Have to do with Documentation Projects?   *267*
   Cathryn Molloy

15 Designing Multimodal Technical Instructions for Cross-Cultural Resonance Using a Culturally Inclusive Approach   *289*
   Audrey G. Bennett

16 How to Write for Global Audiences   *307*
   Birgitta Meex

17 Composing Technical Documents for Localized Usability in the International Context   *327*
   Keshab Raj Acharya

Contributors   *345*

1 Who Is the User? Researching Audiences for Technical Documents

Emma J. Rose

When I think about what makes a good technical document, I'm reminded of Janice (Ginny) Redish's explanation.[1] Redish defines a successful document as one that helps the intended audience find what they need, understand what they find, use that information to accomplish a task and do that in the time and effort they are willing to spend. Makes a lot of sense, right? But how do we learn about the needs of our intended audiences and how can we be sure that the information we produce meets their needs? In this chapter, you will learn about how to research your audience and use that information to help create or improve a technical document.

Technical Documents and Their Users

When you are starting out as a writer, people have lots of advice. If you are a fiction writer, you might get advice like: "Write what you know." When it comes to writing technical information, that advice is dead wrong. Instead, you'll hear "Write for your audience." But you can't just take this advice to heart and then sit in a room by yourself and write. Instead, crafting technical documents always involves others. To learn about others, technical communicators conduct research. You will need to talk with, learn from, and get feedback from the very people who will be using the information that you are creating. At first, this can feel overwhelming, but it's one of my very favorite parts of being a technical communicator.

---

1 This work is licensed under the Creative Commons Attribution-NonCommercial-NoDerivatives 4.0 International License (CC BY-NC-ND 4.0) and is subject to the Writing Spaces Terms of Use. To view a copy of this license, visit http://creativecommons.org/licenses/by-nc-nd/4.0/, email info@creativecommons.org, or send a letter to Creative Commons, PO Box 1866, Mountain View, CA 94042, USA. To view the Writing Spaces Terms of Use, visit http://writingspaces.org/terms-of-use.

A technical document has specialized information to communicate across different genres and modes, and is created by individuals, communities, and organizations, to provide instruction or information. Disparate examples of technical documents include documents, websites, and apps, but also election technologies (Dorpenyo and Agboka), voting policies and procedures (Jones and Williams), hip hop playlists (Hierro), YouTube tutorials (Ledbetter; Edenfield and Ledbetter), or Reddit's Explain it Like I'm Five (Pflugfelder).

Whether we refer to the people as readers, our audience, or users, one thing they share is that they are often "reading to do," meaning they engage with a document to accomplish something (Redish). For this chapter, I use the term *user* to refer to a person who is trying to get something done and has a clear objective in mind. Anytime you approach a new technical writing task, you'll repeat this to yourself: why, what, who, and how which will inform what kind of product or document you will create.

Why: Clarifying Your Purpose

Every technical document has a reason to exist. You need to clarify what the purpose is before you start writing. Having a clear purpose helps to identify who the audience is for your document. Imagine you work for a county health department, and you have been asked to design a document to show people how to properly wash their hands. To get to the heart of the purpose, try the 5 Whys technique (Ries). Ask yourself or your team "Why" five times. Each Why leads to a deeper understanding of the purpose. For example,

1. Why are we designing this document?

    To show people how to properly wash their hands.

2. Why?

    So, people can wash their hands correctly.

3. Why?

    So, they don't get sick or get others sick.

4. Why?

    To help prevent the spread of disease during a global pandemic.

5. Why?

    To save lives.

This is a simple exercise but focusing on purpose helps you connect to the broader implications of a single document and tie it to the broader purpose and mission of the organization.

What: Determining the Kind of Document You Will Create

Different technical documents have different purposes. Some are meant to be references, something that people can quickly look at to get started or complete a task. When you buy a new electronic device, you'll typically see a document called a quick start guide, which is short, includes lots of pictures, and helps you understand the immediate functions of a device, how to power it on, and how to get started. A quick start guide has a focused purpose and is meant to be used early on and then discarded. It doesn't help you learn how to become an expert with your new device. Let's look at another example: a car manual that lives in the glove box of your car. What features of it are dictated by its purpose? Well, most of us don't read it front to back for fun, we also leave it in the car because that is where we will need it. So, it is designed to be small and fit in the glove box. It has an Index, because people reach for it when they have a problem or are stuck, so it needs to help users find different topics quickly and easily.

Technical documents need to support different people with different needs. Take an Electronic Medical Record, which is a software used in many health care settings. The receptionist at a clinic may need to access the scheduling page. They might be interrupted many times but mostly use one or two screens in the system. The nurse may need to primarily access the summary information of a patient that lists their medical history, vitals, and prescriptions. They move from room to room seeing patients quickly and accessing the same screens for different patients. Then, there is the doctor who needs access to reference materials for diagnosing patients and detailed screens to document and capture the nuances of a patient's particular medical conditions. Each of these users is accessing the same technical system, but their relationship to it and their job functions impact their use of the system and their needs for it. Thinking about how technical information is accessed by different users helps us think carefully about who it is we are writing and designing for.

Who: Identifying Your Audience

To gain a deeper understanding of audience, you'll need to gather preliminary information about who it is that you are trying to serve. If you are working within an organization, they may already have a clear idea of who their audience is, but there are nuances and considerations for every technical document. At this stage, it's easy to get overwhelmed with details. So, try and focus on the essential aspects of audience.

We all have assumptions about the people we are writing or designing for. It's helpful first to clarify what you know, or what you think you know, about the audience. Take some time to write down your reflection:

1. Who needs this information?
2. What are the questions they might have?
3. What is their existing knowledge about the subject?
4. Where or how might they access this information?
5. What are the cultural, social, or economic factors that might impact how they access or use this information?
6. What patterns of exclusion exist?

Reflecting on what you know or what you think you know is helpful. However, beginning writers often rely on their intuition or existing experience for understanding audience, which can be risky (Lam and Hannah). When you rely on your own experience, you write from assumptions. One way bias can appear in writing is when you assume other people have similar experiences as you do. There are many examples of forms that ask people to identify their gender and only include choices of men or women, instead of supporting inclusive ideas of gender (Bennett). The people who have designed these forms have a biased view that did not incorporate ideas of non-binary gender expressions. Here is another example: Facebook has a year in review feature designed to celebrate fun memories from the past year. However, user Eric Meyer logged in and was confronted with a picture of his daughter who had died that year (Wachter-Boettcher). The designers of that Facebook feature were biased to think everyone wanted a feature to celebrate their year.

In technical documents, instances of bias arise all the time. I worked at a company that decided to update an older main frame software system to be web-based. The designers of the new system assumed users would appreciate a more modern web-based tool. What they didn't anticipate was that this older system supported very fast data entry for repetitive tasks. The new system, while well designed, required users to click through multiple pages to enter data and it slowed down their work considerably. The designers in that case did not consider how people using this system valued efficiency over other considerations like ease of use.

The best way to prevent bias is to not just think about audience but to meaningfully include users during the design process. Most work in technical communication is collaborative. Ensuring you have a diverse group of people working together is a good start. People who have different lived experiences and perspectives bring valuable expertise to the design process. Further, to ensure your design is meeting the needs of the people who will use it, you need to go further and get to know who the people are who will use your document through research. Through your research, include a range of diverse people in the process and recognize and address exclusion. One approach is to work with and include people who are exclusion experts (Holmes). An exclusion expert is a person who has in-depth, personal knowledge of what it means to be left out. You need include a community of exclusion experts, people who might be disadvantaged by your product or service and ensure your work does not exclude or marginalize them.

The only way to make sure that something that you have designed will work is by getting input and feedback from users along the way. Audience analysis is an ongoing and iterative process. Users and their needs change over time, so it is important to continuously learn about the people you are designing for. The job of understanding your audience is never done.

How: Techniques for Researching and Understanding Audience

There are many ways to learn about your audience. We often refer to these ways of understanding audience as user research methods. Redish uses a metaphor of a toolkit when it comes to understanding users (Redish). You have a toolkit that contains lots of different tools, but not every tool is right for the job. As a beginning technical communicator, your job is to learn about the tools you can use to understand your audience and to learn what each tool is good for and how and when to use it. In the following section, I'll explore a few examples of understanding users but there are many more. To learn more about many of the methods you can use to learn about your audience, check out the Usability Body of Knowledge (https://www.usabilitybok.org/methods) or IDEO's Design Kit (http://www.designkit.org/methods.html).

Examining Existing Forms of Data

If you are working in an organization, they probably already have some information about who the users are. Some of this might be in the form of research and it might be based on people's assumptions. Also, even if there is an existing understanding of your audience, it is helpful to remember that people and contexts change over time. Start with existing data and go further. Here are a variety of methods to help you work with existing forms of data.

- **Interview stakeholders and internal experts** conduct interviews with stakeholders, or people who work inside your organization. This could include product managers or owners, marketing professionals, or people in leadership. Interviewing stakeholders who have deep knowledge of the organization and its goals can help you understand some of the whys related to the document. Additionally, consider interviewing people who have direct interactions with the audience. People who work on the frontlines with your audience such as customer service, helpdesk staff, or people who work in the field. These internal experts tend to be the most knowledgeable about the audience and their needs.
- **Review internal data:** Most organizations have existing data from surveys or market research. There may be reports on customers or people you are designing for. During your stakeholder interviews, ask around and see what additional reports or data might be available that can inform who the audience is. If you are re-designing a website, you can review data analytics which can show you how people currently use the existing site: which pages are the most popular and which search terms people use on the site.
- **Social media:** There are lots of ways to use social media to learn more about your audience. If you are internal to an organization, you can look at Twitter Audience Insights, LinkedIn, or Facebook Audience Insights to learn more about your users' interests, geographies, job titles, demographics, and so on.

These are just a few of the methods you can use to research your audience. They are an effective way to gain an initial understanding of your audience because they leverage information that may already be available. Looking at these existing sources can be a good way to get started familiarizing yourself with your audience's needs, tasks, and concerns.

GATHERING NEW DATA THROUGH RESEARCH

If you've learned all you can about your users from internal sources or don't have access to that information, it's time to directly interact with users to gather more information about who they are and what they need. There are many different research methods you can choose from. Here I will focus on a few of the most common. These methods are especially helpful early in the process when you are still trying to decide who your users are and what their goals, needs, and motivations are.

For any process where you are gathering information, it is important to follow strong ethical standards and guidelines. If you are working for an organization, you should first check with an Institutional Review Board, the legal department, or other people conducting research to understand how to conduct the research in an ethical and just way. Regardless, you should follow these three basic principles:

- **Research is voluntary.** People should not feel pressured to participate, they can choose to participate or not, and they will not be penalized in any way for not participating. They also should be able to change their mind at any point.
- **Purpose, transparency, and use of data.** Participants in your study should be informed of the purpose of the study. You should also tell them what data will be collected, why it is being collected, and how it will be stored and used. Participants should be given the right to remove their data from the study at any time.
- **Consent and clarity.** All aspects of the study and data should be explained in a clear and plain way to participants. Participants have to give their permission, also called consent, to take part in the research. Participants can give their consent verbally or they can sign their name as a way to provide consent. Participants can withdraw their consent at any time.

These factors are particularly important to researching audiences because as a researcher, you have an obligation to conduct the research in an ethical way that protects the rights and privacy of the people who agree to participate. In my own professional practice, I have had several experiences that have required me to take an individual ethical stance. One organization I was working with wanted to use some of the research results, specifically positive quotes from customers, in their marketing materials. This was clearly outside of the bounds of what participants agreed to. I had to deny this request and explain why that wasn't appropriate and violated trust with participants. While I prevailed in this case, acting ethically requires the actions of individuals, but more importantly policies, guidelines, and a coalition of other people, including leaders, to ensure that we are upholding high ethical standards when it comes to research.

If you are interested in learning more about research ethics in designing technical documents, there are many helpful resources to read that examine this topic in more detail, including Victor Yocco's article "Ethical

Considerations in UX Research" and Alba Villamil's "The Ethical Researcher's Checklist."

In the following sections, we'll look at some specific types of research methods you can use to learn more about the people who you are designing for.

Surveys

Surveys are a good way to gather information from many people in a short amount of time. A survey can help you understand who your users are and what they are trying to do. It can also help identify their attitudes and opinions. When trying to understand your audience, it is helpful to use surveys at different times in the design process. For example, if you are designing something new, or embarking on a redesign, it is helpful to survey users at the very beginning of the project. If you are trying to get a sense of how users are experiencing your product over time or compare results or changes, it is helpful to conduct a survey multiple times, such as before a redesign and after a redesign to compare results.

In terms of how many people to include in a survey, there are two key factors to consider: representativeness and sample size (Sauro). The first factor is *representativeness,* which means that the people you are surveying are representative of the people who you are interested in learning from. For example, if we were trying to understand how students experienced a college website, we could feel confident in our survey if we sent it via a student email list and if we added a question asking respondents if they were students at the institution. We would feel less confident if we just posted the survey link to the website because all kinds of people access a college website: staff, faculty, community members, and so on. The second factor is *sample size,* which looks at how precise your results are compared to the broader population. A sample size of 100%, meaning every person in the population filled out the survey, would be highly unlikely. Working from the previous example, it would be surprising and highly unlikely to have every student at a college fill out a survey about the website. Instead, sample size is really a question of how precise you need to be. The more precise you need to be, the larger sample you will need. Statistician and author Jeff Sauro provides a helpful explanation of sample size by asking yourself what margin of error you are comfortable with and working backwards from there by using confidence intervals. For example, Sauro shows the difference in sample size by showing a table that compares numbers based on what margin of error you are comfortable with (see the full chart at https://measuringu.com/survey-sample-size/). If you are asking a question about a rating scale and want to achieve a 1%

margin of error, you would need to survey over 6000 people. But on the other end of the spectrum, you are comfortable with a 20% margin of error, you only need to survey 18 people.

On a final note about surveys and sample sizes, keep in mind that the kinds of surveys we do in writing and designing are different than those that are used for political polling. The precision is often less important in the work we do and a higher margin of error is acceptable. You should think of surveys as one of the many tools you have in your toolkit that you might use to understand people. Using multiple methods (surveys, interviews, and usability studies) can also help you feel more confident in your knowledge and understanding of the people you are designing for.

Surveys are typically delivered online via email or on a website and can include both quantitative and qualitative questions. A survey can ask questions about a person's role, what task they are trying to accomplish, and how they might rate a current product or service. There are lots of helpful tips for designing and writing surveys (Feinstein). For example, you should with general questions and then move to more detailed ones. It's helpful to write clearly and plainly and provide options that don't overlap. Also, keep it short to ensure more people will complete it. It's important to make sure your survey is clear, so have someone try it out before you send it out more broadly.

Let's imagine we are helping a transit agency design a new website and we want to understand our users. Some questions you could ask in a survey include:

1. How often do you use public transportation?
    - o   Everyday
    - o   Several times a week
    - o   Several times a month
    - o   Less than once a month
    - o   Never
2. What are your motivations for using public transportation? (Check all that apply)
    - o   It's my only form of transportation
    - o   To save money
    - o   To avoid traffic
    - o   To relax on my commute
    - o   Other _____
3. When you visited our website today, what information were you looking for? (Check all that apply)
    - o   Schedules
    - o   Fares

- Service delays
- Careers
- News
- Other _____

4. How would you rate our current website?
   - Excellent
   - Very good
   - Good
   - Fair
   - Poor
5. What is your biggest frustration or concern with our existing website? (Open-ended)

Note that, in this example, you are providing individuals with a range of answers to select from (closed questions) vs. allowing them to respond with any response they wish to use (open questions). There are benefits to both. Closed questions give the survey respondents a predefined list which often makes it easy to quickly choose between and provides the data in a form that is easy to compare. Open questions can take a bit longer to complete but allow respondents to use their own words to describe their responses. However, for the researcher, open questions take longer to analyze. It's a good idea to use a mix and be thoughtful about the range of options you are giving a respondent to choose between.

There are a variety of tools you can use to design and distribute a survey. You can send a link to a survey through email or post it on a website or promote it through social media. Keep in mind that how you recruit people might influence your results, so try and be mindful of where you are posting it. Most digital survey tools provide analytics to show you how people are accessing the survey. When your results come back, you can analyze the data and look for patterns. It can be helpful to filter the responses to uncover new patterns. For example, if we looked at the prior survey and focused on people who use transit every day, they might have different needs, tasks, and concerns than people who use it less than once a month. That information can help inform who your audience is and what information they might need.

INTERVIEWS

Interviews are a valuable and informative way to learn more about your audience. Interviews can be quick and informal: you can approach people in public spaces and ask a few questions in five minutes. Alternately, they can be more in-depth, where you recruit ahead of time and spend 30-60 minutes asking questions. Interviews are conducted one-on-one either in person or remotely. For an interview, you should have a clear list of open-ended questions that helps to start a conversation with a representative member of your audience. Your questions should focus on who they are and what they want to do.

Take for example, this list of questions, developed by Derek Ross to better understand audiences for environment-related communication (Ross). These interviews were designed to be conducted at public sites that drew a lot of visitors.

Questions:
- Where are you visiting from?
- What brought you out here today?
- How did you hear about (dam name/park name)?
- Now, just to change up a bit, when I say the word "environmental," what is the first thing that comes to mind?
- Where do you stand on environmental issues?
- Based on that, where do you feel that you get your values from? Do you think of yourself as an "environmentalist," "environmentally active"? Could you explain why or why not?
- Can you think of any environmental arguments that you have heard that are particularly effective?
- Based on what you were just saying, what kind of arguments would be most effective, at least for you personally?
- Looking out over all of this, how would you describe this to someone else?
- And, last but not least, do you mind if I ask how old you are?

As opposed to the closed survey questions we examined earlier, these interview questions are more open. Individuals can respond to the questions using their own words rather than selecting from an existing set of possible responses. Using open questions in interviews is helpful because it provides participants with the ability to choose their own words and language to respond. It also allows the interview to be more conversational in nature which helps create a good rapport between the researcher and the participant.

In terms of how many people you should interview when researching audience, there isn't a clear consensus and it depends on several factors. According to Donna Bonde, you should take into account the following factors: scope of your study, characteristics of the target audience, expertise of the researchers, resources (budget and time) and research audience, and complexity of the project (Bonde). However, a good guideline is to interview between 5-12 users to reach saturation. Saturation is the term researchers use to identify when you are starting to learn the same things from new participants.

Finally, interviews can be done throughout the writing and design process but are usually conducted early in a project to help you, the researcher, understand the needs of the audience and their existing knowledge. Learning this from users will help you make decisions about what information to include, how to design the information, and what features should be included.

To develop your own interview questions, first think about what you want to learn or clarify about your audience. Focus specifically on information that would be helpful as you make design decisions. Interviews help gather people's opinions, preferences, ideas, and opinions in their own words and language. However, they also have drawbacks. Interviews rely on participants' ability to accurately recall details and they may have a desire to respond to a question in a way that meets the expectations of the person conducting the interview.

For example, I live near Seattle and recycling is a badge of honor. If I wanted to ask people about their recycling habits, I could interview them and I'm sure I would learn a lot about their opinions, beliefs, and preferences around recycling. But if I wanted to understand their behaviors, actions, or habits, then I would need to engage in observational methods and watch how they recycle and what choices they make. What they tell me about their recycling habits might be different from what I observe. The next method, contextual inquiry, is an example of an observational method.

Contextual Inquiry

Similar to an interview, a contextual inquiry is a method where a researcher asks a participant questions (Beyer and Holtzblatt). However, there are some important differences. A contextual inquiry is conducted at the site where the technical information is being used. For example, if we wanted to understand how to design a new billing system for a doctor's office, we would conduct the contextual inquiry at the doctor's office with the person who does the billing while they were doing it. A contextual inquiry includes questions about a person and their preferences, but it also includes watching a person work with an existing system to understand how they do their work and what challenges they face. When doing a contextual inquiry, the interviewer puts themself in the mindset of an apprentice, trying to learn and understand what a skilled expert does.

You would use this approach to research your audience when you are trying to learn how people currently approach and accomplish a task. It is especially helpful when you are approaching a topic you know very little about or the topic is highly technical or complex in nature. Contextual inquiry also helps show you how people get things done rather than just their recollection or how they do something. Surveys and interviews are helpful to understanding people opinions and their recollections of a task, but contextual inquiry is beneficial to give you a deeper understanding and helps to highlight practices that people might not remember or take for granted. Similar to interviews, there aren't strict guidelines of how many people you should include in a contextual inquiry, but a good rule of thumb is to start with five participants from each unique user group. If you are still learning new things after talking to five people, add five more.

Synthesizing: Capturing What You Learn about Your Audience

When you research your audience, it's helpful to synthesize that information and capture it in a way that helps you articulate who your audience is for your own design and writing practice. It is also helpful for you to share that knowledge with others on your team or in your organization. All the methods described earlier give you different insights into your audience. So, to get a better picture we often engage in what is called triangulation. Triangulation refers to using different methods to understand a single phenomenon (Denzin). Each method provides some insight, but taken together, they can help fill in gaps or shine light on certain aspects of who your users are. Looking across the data from multiple methods, you can capture key insights about your audience and start to get a picture of who they are and what they need.

For example, let's imagine we are redesigning the website for a transit organization, we'll call it Big Bus Org. We may choose to employ several research methods. For data, we could look at existing data to understand how people use the current site, what are the most popular pages, and what search terms do they use. We could also interview customer service representatives and ask them what are the common questions they get when people email or call. We can also look at the social media accounts for Big Bus Org and see what prompts people to reach out for help and when they are frustrated. We could also conduct new research to better understand Big Bus Org customers. We could post a survey on the home page of their existing website to understand people's needs, tasks, and how they currently rate the site. We could also interview existing customers and schedule these interviews over the phone or remotely to ask people about their experiences with the website and riding the bus. We could also do contextual inquiry by asking people if we could join them on a bus ride or watch them navigate the existing website looking for information.

After collecting this data, we could then organize it in several ways. We could first catalog all the complaints, concerns, and frustrations of the current web site experience, which can help us develop requirements for our new design. We can also synthesize the data from the sources and try and group and conceptualize our audience so we can understand who they are and communicate about our users to others on the team.

Personas are a useful tool for capturing data about your audience's needs, goals, and motivation. You take what you learn across multiple methods and come up with a composite, abstract, portrayal of an imaginary person. You might have 3-5 different personas that represent different aspects of your audience. If we take the example of a transit rider from earlier, we might have three personas: Daily choice rider (someone who has a car, but chooses to commute daily), Transit dependent rider (someone whose doesn't own a car and relies on transit), and an Event rider (someone who rides infrequently because of a special event like a concert). Each of the three personas have different motivations, familiarity with the system, and needs. Segmenting your audiences in this way can be helpful for making decisions about how to design your technical information.

A persona has two functions: empirical, in that they capture data, and rhetorical, in that they communicate that data to others (Rose and Tenenberg). Once you capture and synthesize your data, you are ready to create a compelling and persuasive communication product that can help educate others about the audience. Personas typically contain information about one fictional user who represents many other users. It can contain a photograph of the person, a summary of who they are and what they are trying to accomplish with your product or service. A persona can also include an emblematic quote from that user and demographic or psychographic information. If you want to learn more, there are lots of excellent resources that can help you learn more about creating personas (Adlin and Pruitt; Mulder and Yaar).

Creating a persona helps you synthesize the information to better understand and articulate the needs of your target audience and communicate who they are to others on the team. They are also helpful when you are drafting, editing, and revising your technical document. Let's go back to our example of a transportation website. For frequent bus riders, here is Craig, a bus rider who owns a car but prefers to take public transportation due to concerns about the environment and wanting to relax on his commute. The second persona is Louise, she is dependent on public transit to get around as it's her only form of transportation and she is highly knowledgeable about routes and services. As you start working on your website, you might ask: "What information does Craig need to plan a trip?" or "What does Louise need to know about purchasing a monthly pass?" Imagining the specific people who will use your information will help you keep the focus on their needs and avoid thinking of users generically. Alan Cooper refers to this as "the elastic user" (Cooper) when you have a team sitting around thinking of users as generic abstractions, they can stretch to fit any scenarios or situation the design team can think of. Use your personas to focus on specific needs and attributes and communicate those to others on your team. This helps avoid the problem of the elastic user.

Personas also often include details about the user's context or setting that helps make decisions about design. For example, we may have learned through research that Louise's persona accesses the website on a mobile device and has a limited data plan. This will help remind the design team that there needs to be an optimized mobile version of the website that is lightweight and runs quickly. We might also learn that Louise is busy and in a hurry in the morning and so the schedule page should provide information that is easy to read which can help designers think about how to format that information.

Evaluating Your Technical Documents: Usability Testing

Up to this point, we have discussed how to research your audience to help inform the writing and design of technical information. However, if you want to ensure what you have designed works for your readers or users, you also need to conduct usability testing. Usability testing is a method that helps you understand if a person can successfully use what you have designed and what problems arise when they try to use it. Usability testing can and should be used throughout the design process. That means, don't wait until you are done. Gather input by showing early drafts and prototypes to users early in the process. That way it's easier to make changes, decisions, and adjustments as you go. According to Redish, there are six characteristics that all usability tests share:

- **Real issues:** Thinking about what you want to learn, plan your test accordingly.
- **Real people:** People participating in the study who represent the audience you are designing for.

- **Real tasks:** Different stories or scenarios that are realistic for your audience that they will do when using your product.
- **Real data:** Information you collect in the study, when you ask questions, watch, listen, and take notes.
- **Real insights:** Hold off on your assumptions, review the data, and determine what is working and what needs to be improved.
- **Real changes:** Take what you have learned and make meaningful changes.

In the next section, we'll look at an example of an informal usability study you could conduct.

Usability Testing Examples

Let's look at a usability testing example that you could use to practice. I have also developed a template for a Usability Testing Plan that you can use when planning and conducting your own study. Visit https://doi.org/10.1145/3548658 and look for the file in the Educational Resources section (Rose, 2023). We'll continue to work on the idea of improving a website about public transportation. For this example, you'll need three people, a computer or mobile phone with access to the internet, your local transit organization. If you don't have one locally, try the closest big city or the Amtrak website. It is also helpful if you can record the session, so you can go back and review what happened. If you are meeting remotely, a tool like Zoom can let you share screens and record the session. If you are meeting in person, you can use a video camera or mobile phone to record. In the following section, you will find an example usability test session, from http://www.usability.gov.

Example Usability Test Session

Here is an example test session.

1. The facilitator will welcome the participant and explain the test session, ask the participant to sign the release form, and ask any pre-test or demographic questions.
2. The facilitator explains thinking aloud and asks if the participant has any additional questions. The facilitator explains where to start.
3. The participant reads the task scenario aloud and begins working on the scenario while they think aloud.
4. The note-takers take notes of the participant's behaviors, comments, errors, and completion (success or failure) on each task.
5. The session continues until all task scenarios are completed or time allotted has elapsed.

6. The facilitator either asks the end-of session subjective questions or sends them to an online survey, thanks the participant, gives the participant the agreed-on incentive, and escorts them from the testing environment.

Example Scenarios

For the study on the transit website, here are five scenarios you can use.

1. Let's imagine you are commuting to school or work. You need to find a way to arrive by 8am next Monday morning. Using this website, find out what route will get you to your destination by 8am.
2. You have a friend visiting from out of town. Find out if there is a route from the airport to your home.
3. You want to buy a monthly pass to use public transportation. Find out how much a monthly pass costs and where you can purchase one.
4. You think your bus or train is delayed. Find out if there are any delays in your area.
5. You have heard the transit agency is thinking about cutting services in your area. Find a way to provide feedback to the agency and voice your concerns about cutting services.

Analyzing and Reporting Data

After conducting a usability test, the team needs to make sense of the data. Look through the data you have collected, read your notes, and look for patterns or areas where participants struggled. The goal of a usability test is to find problems, so look for examples of problems that people encountered. You may have captured quantitative data such as success rates, task time, error rates, and questionnaire ratings. You have also collected qualitative data about what participants did, what they said, comments, recommendations, or frustrations. For each of the problems you find, make a list of the Finding (what was the problem) and the Evidence (how do you know it was a problem). Evidence can include quantitative and qualitative data, using quotes from participants is a good idea to help demonstrate the problem. For example:

Finding: Users struggle to find information about service delays.

Evidence: Three out of five participants were not able to successfully find the service delay page. The remaining two participants found the page but stated confusion about the information on the page. Participant

3 said "I think I'm in the right place, but I'm just not sure if this answers my question."

After you have identified your top findings on the site, you can communicate this information in a written report or oral presentation to the rest of your team. It is helpful to include screenshots of the document or information and include quotes or video clips that show participants' experiences. If you have a large number of findings, consider assigning a severity rating to each one, for example: critical, serious, or minor.

Making Changes and Improvements

Once you have gained a clear understanding of what is working well and what needs to be improved, it is time to make the changes to the original technical document. Take what you've learned from the test and make improvements in a new iteration. Prioritize the most critical problems and fix those first. Ideally, you should test your document again to make sure the changes have improved the document overall.

In this section, we have just touched on the method of usability testing. There is a lot more to learn and many helpful resources to do so, including how to books like *Usability Testing Essentials* by Carol Barnum and *Don't Make Me Think* by Steve Krug. You might also consult access resources available on the website of the User Experience Professionals Association (https://uxpa.org) or Digital.gov (https://digital.gov).

Concluding with Some Tips for Research

As you have now learned, there are many ways to understand and research your audience for the technical document or design you are creating. It can feel overwhelming when you are starting out, so I'll conclude with some tips:

1. **Identify and then check your assumptions:** You probably are going to over-estimate what you know about your audience. It will be important to separate out what you think you know about your audience, your assumptions, and what you can back up or verify with data.
2. **Recruit for breadth and depth:** Find participants to give you feedback on your product who are target users, but also be sure that the people who you talk with make up the diversity of the population. Make sure to seek out people from marginalized or underrepresented audiences. To create inclusive technical information, you should engage exclusion experts.

3. **Start small:** As you begin to learn about your audience, start small. Choose one method or one group of users to learn more about. Then build up your knowledge over time by adding additional methods and broader groups.
4. **Follow an iterative process:** Involve representative users early on and throughout the process to ensure you are creating technical documents that meet their needs.

One thing is certain, it's never too early to start learning about your audience. So, try out these methods and learn about new ones so you can create helpful and successful technical documents.

Work Cited

Adlin, Tamara, and John Pruitt. *The Persona Lifecycle.* Morgan Kaufman, 2006.

Barnum, Carol M. *Usability Testing Essentials: Ready, Set-- Test.* Morgan Kaufmann, 2011.

Bennett, Micah. "Beyond the Binary: 5 Steps to Designing Gender Inclusive Fields in Your Product." *UX Collective*, 2020, https://uxdesign.cc/beyond-the-binary-5-steps-to-designing-gender-inclusive-fields-in-your-product-ff9230337b4f.

Beyer, Hugh, and Karen Holtzblatt. *Contextual Design: Defining Customer-Oriented Systems.* Morgan Kaufmann, 1997.

Cooper, Alan. *The Inmates Are Running the Asylum.* Sams Publishing, 1999.

Denzin, Norman K. "Triangulation." *The Blackwell Encyclopedia of Sociology*, Wiley Online Library, 2007.

Dorpenyo, Isidore, and Godwin Agboka. "Election Technologies, Technical Communication, and Civic Engagement." *Technical Communication*, vol. 65, no. 4, 2018, pp. 349–52.

Edenfield, Avery C., and Lehua Ledbetter. "Tactical Technical Communication in Communities: Legitimizing Community-Created User-Generated Instructions." *SIGDOC 2019—Proceedings of the 37th ACM International Conference on the Design of Communication*, 2019, https://doi.org/10.1145/3328020.3353927.

Feinstein, Tanya. "Writing Usable Survey Questions: 10 Things All UX Researchers Should Know." *UXPA Magazine*, vol. 17, no. 5, 2017, https://uxpamagazine.org/writing-usable-survey-questions.

Hierro, Victor Del. "DJs, Playlists, and Community: Imagining Communication Design through Hip Hop DJs, Playlists, and Community: Imagining Communication Design through Hip Hop." *Communication Design Quarterly*, vol. 7, no. 2, 2019, pp. 28–39, https://doi.org/10.1145/3358931.3358936

Holmes, Kat. *Mismatch: How inclusion shapes design*. MIT Press, 2020. https://doi.org/10.7551/mitpress/11647.001.0001.

Jones, Natasha N., and Miriam F. Williams. "Technologies of Disenfranchisement: Literacy Tests and Black Voters in the US from 1890 to 1965." *Technical Communication*, vol. 65, no. 4, Nov. 2018, pp. 371-386.

Krug, Steve. *Rocket Surgery Made Easy: The Do-It-Yourself Guide to Finding and Fixing Usability Problems*, 1st ed., New Riders Publishing, 2009.

Lam, Chris, and Mark A. Hannah. "Flipping the Audience Script: An Activity That Integrates Research and Audience Analysis." *Business and Professional Communication Quarterly*, vol. 79, no. 1, 2016, pp. 28–53, https://doi.org/10.1177/2329490615593372.

Ledbetter, Lehua. "The Rhetorical Work of YouTube's Beauty Community: Relationship- and Identity-Building in User-Created Procedural Discourse." *Technical Communication Quarterly*, vol. 27, no. 4, Routledge, 2018, pp. 287–99, https://doi.org/10.1080/10572252.2018.1518950.

Mulder, Steve, and Ziv Yaar. *The User Is Always Right*. New Riders, 2006.

Pflugfelder, Ehren Helmut. "Reddit's 'Explain Like I'm Five': Technical Descriptions in the Wild." *Technical Communication Quarterly*, vol. 26, no. 1, Routledge, 2017, pp. 25–41, https://doi.org/10.1080/10572252.2016.1257741.

Redish, Janice. "Letting Go of the Words." *Letting Go of the Words*, 2nd ed., Morgan Kaufmann, 2012, https://doi.org/10.1016/C2010-0-67091-X.

Ries, Eric. "The Lean Startup: How Today's Entrepreneurs Use Continuous Innovation to Create Radically Successful Businesses." *The Lean Startup*, Crown/Archetype, 2011.

Rose, Emma J., "Usability Testing Plan Template: A Flexible Tool For Planning and Teaching Usability Evaluation." *EngageCSEdu*, vol 1, no. 2, 2023. https://doi.org/10.1145/3548658.

Rose, Emma J., and Josh Tenenberg. "Poor Poor Dumb Mouths, and Bid Them Speak for Me: Theorizing the Use of Personas in Practice." *Technical Communication Quarterly*, vol. 27, no. 2, 2018, pp. 161–74, https://doi.org/10.1080/10572252.2017.1386005.

Ross, Derek G. "Deep Audience Analysis: A Proposed Method for Analyzing Audiences for Environment-Related Communication." *Technical Communication*, vol. 60, no. 2, 2013, pp. 94–117.

Sauro, Jeff. "What Is a Represenbtative Sample Size for a Survey?" *Measuring U Web Site*, 2010, https://measuringu.com/survey-sample-size.

Villamil, Alba. *The Ethical Researcher's Checklist*. UXR Conference, 27 June 2020. https://joinlearners.com/talk/the-ethical-researchers-checklist.

Wachter-Boettcher, S. *Technically Wrong: Sexist Apps, Biased Algorithms, and Other Threats of Toxic Tech*. E-book ed., W. W. Norton, 2017.

Yocco, Victor. "Ethical Considerations In UX Research: The Need For Training And Review." *Smashing Magazine*, 2020, https://www.smashingmagazine.com/2020/12/ethical-considerations-ux-research/.

Teacher Resources

Overview and Teaching Strategies

Researching audiences can be addressed at several points in a technical communication course. Any document that is being written or designed for an audience beyond the instructor would benefit from activities related to researching audiences. For most assignments in technical communication, the 5 Whys exercise is helpful for students to understand and articulate the exigency for crafting a document. Further, for most assignments, it is helpful to have a research and evaluation phase and to choose methods accordingly.

In the research phase, students can explore how to understand audience by looking at existing data or collecting new data, as explained in the previous chapter. Some examples of assignments and corresponding activities are:

1. **Resume and cover letters:** Interviews with hiring managers within a specific field to understand what they look for and how they read a resume.
2. **Writing for the web:** Looking at social media or existing data about current usage to understand issues or gaps.
3. **Proposal writing:** Conducting a survey to better understand the existing needs of a community or audience.
4. **Instructions:** Conducting a contextual inquiry to learn the steps and technical know-how of how a skilled user does something.

In the evaluation phase, students can conduct usability testing with a think aloud protocol on most documents and artifacts produced in class. The challenge is to access participants who are representative of the target audience. For educational purposes, you can have students conduct usability tests on each others' work, but having representative users is beneficial if possible. Here are some ideas on assignments and how to conduct usability tests:

1. Websites or instructions – create scenarios to identify problems and ask representative users to perform them on a site or page.
2. Proposal writing – ask a professional or a funder to read through an abstract or proposal to identify strengths and weaknesses.

A key item of researching audience is to help students connect what they learned through research to the decisions they have made in their design or document. One strategy to help students make these connections is to include a reflective memo as an addendum to each assignment. Ask students to reflect on one re more of the following:

1. How does the design of the document meet the needs of users?
2. What specific choices did you make in the design or writing of the document to meet users' needs?
3. How are you ensuring that your document or design is not biased? How does it reflect the needs of the diversity of your audience?
4. What aspects of the document or its design are you unsure about? What areas do you need to gather feedback on?
5. What changes did you make to the document or design based on usability testing or feedback?

Finally, being able to have students experience both the research and evaluation phase is beneficial, it is helpful to provide them with an end-to-end design experience which may span the entire term. Clients, campus partners, and community-based organizations are helpful to partner with. For example, partnering with local business to redesign a website, writing a grant for a non-profit, or working with iFixit.com where students can design and usability test instructions for repairing electronic devices. When working on a larger project such as the ones described previously, having students work in teams can also help share the work and strengthen human skills like communication and collaboration.

Discussion Questions

To help students understand the ideas related to researching audience, consider posing the following discussion questions or activities. Some of these can be explored individually, others are better done in small groups or by a full class.

1. Think of something you have used or read that you felt like wasn't designed for you and you felt excluded. What was it? How did it make you feel? How could you redesign that item, document, or design to make it feel like it was designed for you?
2. Let's imagine we work at a company and we're going to design a new backpack. First, brainstorm all the possible potential users of a backpack. Second, list out all the unique needs of the potential users. Third, in small groups, each group should select one user to design for and list all the features of the backpack they are designing for that one user. Finally, come back together to share out different design ideas. How did the unique group you chose to design for impact the design and features you chose for the backpack your team designed?

3. How are technical documents different from other kinds of documents? Why is this difference important and how does it inform the process we use to design technical documents?
4. In the section that describes the techniques for researching and understanding audience, define each of the techniques and list what are the pros and cons of each method?
5. In the section on usability testing, there is an activity that asks you to conduct your own test. In teams, conduct the usability test on a local transit website. Share your responses to the following questions: What was it like to be in your role (participant, facilitator, notetaker)? What did you learn about the website? What works well? What ideas do you have for changing the site?

2 Assessing Sources for Technical Communication Research

Therese I. Pennell

As academic writers, your technical communication research papers will require that you use reliable sources to understand your topic and support your argument.[1] This chapter provides you with tips on assessing sources to complete these types of papers. In high school you learned how to search for sources, parse out websites based on your domain anchors (.com versus .org, versus .gov), and streamline your work based on ideas from sources. These are important strategies, but your technical communication research papers require sources that are accurate, timely, relevant, focused, and rigorously documented. In college you have greater access to more academic resources, and at this level your technical communication documents require more demanding research standards. You are tasked with assessing sources to meet these standards.

This chapter helps identify the multiplicity of sources you can use, categorizes them based on the rigor of publication, and helps you understand when and how to incorporate one or more sources within your paper to support your claim. You will also find information on using rhetorical reading strategies to help determine a source's credibility and relevance. This chapter provides examples that illustrate how to begin the research process and how to use valid and accurate information to meet the requirements for technical communication research.

So, you're going to do research. Research in college can be an intimidating process. The assignments require deep investigation, and the standards can seem vague or else perplexing. You will come across

---

[1] This work is licensed under the Creative Commons Attribution-NonCommercial-NoDerivatives 4.0 International License (CC BY-NC-ND 4.0) and is subject to the Writing Spaces Terms of Use. To view a copy of this license, visit http://creativecommons.org/licenses/by-nc-nd/4.0/, email info@creativecommons.org, or send a letter to Creative Commons, PO Box 1866, Mountain View, CA 94042, USA. To view the Writing Spaces Terms of Use, visit http://writingspaces.org/terms-of-use.

terms you may not be familiar with and tools you have never used before to complete your research-based assignments. If you complete these assignments citing sources you may have quickly gleaned from a Google search, your instructors may be less than impressed. This chapter will help you in assessing sources for your research. This chapter introduces you to rhetorical reading that provides you with strategies to identify important ideas about the sources you should be using. We look at the different types of sources available to you, and the standards the sources should meet before including them to support your arguments.

Let's begin with the type of research you may complete. Embarking on research can mean you will be doing one of two things or both:

1. Primary research
2. Secondary research

So, what's the difference between these two types of research?

Primary research means that you will go out to find evidence and eventually document it. Some tools used to conduct primary research include: interviews, surveys, and observations. For example, if you had an assignment to determine whether your university's website is user-friendly (i.e., the website is designed to be easy to use, so, as a user you should be able to find information you are looking for with as few clicks as possible), you would observe your peers using the website or ask questions via an interview or a survey. In this case you would have to carry out some background research to help you understand your topic and what to look for in creating your research tool.

In secondary research you use research from authors who have already collected evidence/data and are presenting that information in written format. Usually, your assignments for your technical communication courses will require secondary research. An example of an assignment that requires secondary research is to identify and analyze the five types of memos. You would look up sources that provide information about the different types of memos, find sources that provide samples of each, and analyze parts of each memo. The sources you use for this assignment, as the ones in the planning phase for primary research, require some type of assessment. What follows are some tips about how to find reliable sources that provide accurate information.

In high school you would have learned to do quick web searches. And you may have learned that certain websites were more reliable (and so shared more accurate information) about certain topics. So, a website produced by a corporation or commercial entity (.com) would be less reliable than a website created by a government agency (.gov) on topics cautioning about public health, for example. On the other hand, the corporate website might be more useful to find information about the availability of certain commercial products needed to prevent a public health out-break.

In addition, you should be aware of the source's relationship to the content. In the case of a company presenting their products, as a reader/consumer you should be aware that the company may present a more flattering or biased view of their product. How often do you see advertisements where a product's most advantageous qualities are presented while any disadvantages are conveniently omitted? Commercials where the reach, cost, and quality of a cell phone network are shared, but not that calls are often dropped in rural areas? Similarly, government agencies can present information flattering or biased about some of its policies but neglect to share how it can negatively affect certain groups. For example, enacting stringent immigration policies that they do not reveal contravenes international human rights laws.

The basis of assessing sources, therefore, is that you understand the purpose of your sources and the credentials of the author (in this case, the person or entity that created the website). You are expected to use sources that are credible and reliable for your technical communication research papers. A source that is credible and reliable must be accurate, timely, relevant, focused, and rigorously documented (more on these ideas later).

Rhetorical Reading

You may be wondering at this point, "How will I remember all of this as I am finding my sources?" I must admit that the entire process of assessing sources can seem daunting. But being a technical communication researcher requires a level of responsibility because your research carries weight in your field of study. The process can be made easier using a reading strategy referred to as rhetorical reading. By now in your college career, you may have heard about the rhetorical context, the big three ideas of rhetoric: ethos, pathos, and logos. The rhetorical context, you may know, is an important strategy used to write effectively. And you may ask, "how can it help me with reading?" Aristotle's rhetorical theory was originally to help speakers. Modern researchers adopted it to improve writing strategies, and it is applied to reading as well. Susan K. Miller-Cochran and Rochelle Rodrigo highlight two different types of reading: rhetorical reading and focused reading.

Most of your reading, to this point, may have been finding sources to gather information (Miller-Cochran and Rodrigo 13)- this is focused reading. Rhetorical reading is reading looking at the context of the source. More than just gathering information, you are seeking out:

- **Ethos:** Who is the author, what is the author's background?
- **Logos:** What is the purpose of the source, what is the argument, how does the author make the argument?
- **Pathos:** Who are the intended readers, in what way does the author try to connect to the reader?

Readers using the rhetorical reading strategy read to more than gather information, they look at the author, argument, audience, and purpose of the document. This type of reading allows you, as the researcher investigating the source, an important pathway to assess your source.

At this point, we should explore the type of sources you will be reading rhetorically. Let's start with what you will find at your library, whether you go to your college library physically or log in through its interface. You should first have an idea about what your topic is: keywords or phrases associated with the topic. Think about it like developing a music playlist. It depends on the occasion you use it for— studying, exercising, cleaning, or meditating. How you search for the music depends on what you plug in to search for it. You also would want to have questions associated with the topic— your research question(s). These, the keywords or phrases associated with the research question, can be shared with a librarian in person or typed in the library's interface search bar to help you identify sources.

Sources Categorized

Once you have established your topic and research question(s), next is to determine the type of sources you can use. Let's start by looking at categories of sources/documents you would find through your research.

- Books
- Academic journals
- Conference proceedings
- Periodicals
- Websites/online documents
- Blogs/wikis/self-published online documents

Your college library would grant you access to a wide assortment of resources from where you can begin your research. This amount of information can be overwhelming. But a good place to begin is by understanding each type of source. In the next section, you will find a quick description of each type of sources.

Books

Scholarly books are those where an author or authors present original research on a topic. These books are considered credible sources, that is, well researched. Books, as a scholarly resource, can be broken down into two main categories— books written by subject-matter experts or edited collection books.

Books written by subject-matter experts simply mean the author has done a lot of work in the field of research so the author knows the topic very well. Sometimes these books may have more than one author who contribute equally or almost equally to the content. Books can come under different editions (first, second, third . . . tenth editions), this means that the book is published more than once with updated information.

Another type is the edited collection books. These books have an editor or editors who are often subject-matter experts on the topic, but the book includes works by other authors. So, each chapter of the book may be written by different authors who are experts in the specific sub-topics the book covers. In this case, each of these contributing authors usually writes a chapter on their area of expertise. An example of an edited collection is *Solving Problems in Technical Writing* edited by Lynne Beene and Peter White. This particular book has twelve chapters written by different authors, with each author focusing on a specific topic on which they have expertise. The edited collection can also be in a series, meaning there are multiple books with new and different topics under the umbrella topic. An example of the edited collection in a series is *Writing Spaces: Readings on Writing* published by WAC Clearinghouse and edited by Charles Lowe and Pavel Zemliansky. This edited book in a series includes three volumes, the first looking at academic writing in general, the second on the rhetorical situation, and the last focuses on audience considerations.

Before being published and sold in stores, books go through a publication process. It is important to understand this process to better understand what makes books a reliable source. In general, there are three steps:

1. **The editor:** The writer will have an editor associated with the book publisher who among other things makes sure the book is completed on time. This editor also has the book reviewed by multiple experts in the related area, and these experts assess if the research reported is effective and if the book merits publication.

2. **The copyeditor:** After the writer sends in her/his completed manuscript (the complete and polished draft of the document) the copyeditor will proof-read and edit it. The copyeditor looks at grammar, spelling, punctuation, and sentence style.
3. **The Reviewer:** Once the book has been readied to be published, it is revised for style so that it fits the publisher's style guide.

Each step in the process also contributes to confirming the accuracy of different kinds of information—from content presented to grammar and style used—in the book.

While books are considered scholarly and your instructors will see such a resource as prime research, as an academic researcher it is important to check the ethos of the author, especially of single-author books. Is the author well-known in the field the topic covers? Does the author have other publications? Does the author acknowledge any individual(s) for reviewing the book's facts?

Noteworthy is the publication process of edited collection books, these are the books that have an editor(s) but several authors contribute to it. While the publication process is the same, that is, the editor or editors of the book has a publisher's editor, a copyeditor, and style reviewer; the book editor (who is usually an expert in the field) reviews the content of each chapter and in most cases sends the work out for anonymous reviews. An anonymous review means someone else in the field will read it and give feedback on the content. It is referred to as "anonymous" because the reviewer will not know whose work it is, and the author will not know who reviewed her/his work.

The publication process means that it can take some time before a book is published. As you are aware, also, books are available at a cost which can range to prohibitive prices and this, along with the type of topic and language level used, can limit who can access them. So, the readers of scholarly books can be limited.

As a researcher you will find that scholarly books contain in-depth detail on various topics in technical communication. For new researchers, the context and details are invaluable. An example is *Translation and Localization: A Guide for Technical and Professional Communicators* edited by Bruce Maylath and Kirk St. Amant. This edited collection book allows students to learn tips about writing documents to be translated into other languages, about translation software, errors to avoid, and case studies of translation and localization in global contexts. For the seasoned researcher the specialized focus on topics is indispensable. An example of the edited collection by Kelli Cargile Cook and Keith Grant-Davie, *Online Education: Global Questions, Local Answers* and *Online Education 2.0: Evolving, Adapting, and Reinventing Technical Communication Online Learning* allowed researchers to understand how to apply online learning in the classroom, learn about policies universities implemented for their online programs, and read about practical cases. Their second volume built on these ideas as online technology and strategies shifted and improved.

The details and variety of topics that are covered in scholarly books makes this resource invaluable to technical communication research. Students tasked with researching technical communication topics that requires they be knowledgeable of the subject or an aspect of the subject—for example, understanding what a thing is, the contextual background of that thing, its rhetorical value, and application in practical settings—would do well in using books to complete their research.

Academic Journals

Another important scholarly resource in technical communication research is the academic journal. In every field of study there will be multiple academic journals associated with it. Some examples of journals in the field of technical communication include: *Journal of Technical Writing and Communication* (JTWC), *Programmatic Perspectives, and Technical Communication Quarterly* (TCQ). Academic journals (journals for short) are considered important resources for research. This is where researchers in your field share their work. They are often associated with an academic organization or institution and a publication company. JTWC is a member of Committee on Publication Ethics (COPE), it is associated with Iowa State University, and is published by Sage publishers. Programmatic Perspectives is the journal of Council for Programs in Technical and Scientific Communication (CPTSC) and the journal is published all online, thereby requiring no formal publisher. TCQ is a journal of the Association of Teachers of Technical Writing (ATTW) and is associated with Taylor and Francis publishers.

Similar to books, journals have a review and publication process, which, if you are taking notes, means there is a rigorous approach to publishing the work. As a result, it can sometimes take years before it can be published. In general, this process works as follows:

1. **The editor:** Usually a researcher in the field who is a member of the organization the journal is affiliated with and is in close contact with the publisher. The editor usually does an initial review of a manuscript to determine if the research is effective (per standards in the field) and determines if the manuscript should move to the next stage of review or be rejected (not considered for further review for publication).

2. **The anonymous peer reviewers:** Usually members of the research field who are experts in the topic covered in a manuscript and who volunteer to review manuscripts for the journal. Typically, a writer's manuscript would be sent to two or three anonymous reviewers to obtain their expert opinion on the research presented in the manuscript. This means that the reviewer will not know whose work they are giving feedback on and the writer will not know who gives feedback. If these reviewers find the manuscript has merit and the research is effective, they recommend its publication in the journal. Accepted manuscripts then move on to the next stages of the publication process.
3. **The copyeditor:** Once the manuscripts have been formally accepted to be published, the book editor has a copyeditor edit the article for spelling, grammar, punctuation, and sentence style.
4. **The style reviewer:** Once the document has been readied to be published, a reviewer reads and revises the document so that it fits the publisher's style guide.

The journals would have a standard editor who publishes the journal either twice a year or quarterly (four times a year). These editions are known by seasons: winter, spring, summer, fall editions. For manuscripts, editors send a notice or a call for papers to academic researchers in their field of study to share research on current topics. These manuscripts are then forwarded to anonymous reviewers. The anonymous reviewers read and provide feedback to the book editors to reduce the chance of bias (whether for or against the writer), the editor then forwards feedback to the writer. Quite often the writers are asked to revise and resubmit and then their work will be resubmitted for review. The journal publication process includes a few more iterations than a book.

In addition to the rigor of publication, journals usually require subscription, and either the printed copy is mailed to you, or you get electronic access to the journal articles using login information. The limited access helps determine the audience of journals: whether researchers and/or members of a specific discipline. If you are a member of the organization to which the journal belongs, you pay a fee (dues) covering the cost of access. Non-members are able to pay for subscription to the journal as well. Some "open-access" journals can be shared with the general public for a small fee or free of charge.

As students, the college pays for subscription to many journals, which are accessible using your college login information through your library. Remember, if a journal specific to your field is not available through your library, you can request that the library buys subscription to the journal or you can request an inter-library loan (ILL) to get access to specific articles through libraries that have subscriptions to the journals.

Journals are an invaluable resource to technical communication researchers because they provide focused and in-depth information on a topic. A journal article provides extensive information on past research completed (review of literature), description of the researcher's methods, discussion, and future research ideas on the topic. Your professors often may ask that you use scholarly sources like journal articles in your research papers. This is helpful given that non-scholarly sources sometimes give generalized information that is best supported with the research provided in journal articles. From journal articles, you get the context, the research methods, and findings directly from the researcher. For this reason, journal articles are best used when you need to inform yourself on a topic and when you need to support your research with information from a credible source.

Conference Proceedings Papers

We have spoken a little about technical communication organizations (CPTSC, ATTW). Now, take for example, CPTSC, a technical communication organization that produces the journal, *Programmatic Perspectives*. CPTSC also hosts annual conferences. These conferences list topics as foci of discussion, and academic researchers are asked to add to that discussion by sharing their experiences, on-going research, completed research, and/or research findings. These conference discussions have a format: a moderator introduces the topic and presenter(s), an individual or panel of speakers present, and their presentations are followed by a question and answer (Q&A) session. While the conference discussions are carried out orally, conference proceedings are a written, more formal version of the presentations, often adopted based on feedback from the Q&A session.

Conference proceedings can be published by the organization that coordinated the conference either via its affiliated journal (CPTSC would publish their annual conference proceedings in their journal, *Programmatic Perspectives*) or via a book series or special serial edition. Conference proceedings share similar publication process as edited books: there is an editor affiliated with the publishing company and generally also a copyeditor and style reviewer. Another similarity is that the editor may review or have anonymous reviewers give feedback on manuscripts before publishing. What makes conference proceedings different from journal articles and edited books is the contents of the proceedings are written versions of the presentations given at a particular conference. (Participation in most journal and edited book projects tends to be open to any author to submit a manuscript for publication consideration with them.)

Proceedings are also usually accessed in print or electronic formats with a subscription. Generally, members of the organization that coordinated the conference receive access to proceedings as part of their membership in the organization. Individuals that are not members usually have to purchase a subscription to access proceedings content (often called "papers") or buy individual papers published in a proceedings. The limited access to conference proceedings can help us determine the audience of these documents: researchers and members of the specific field of discipline. Conference proceedings papers are another helpful resource you can use for your research.

Conference proceedings papers offer focused research on a specific topic, the more recent ones usually feature current topics, making them another invaluable resource in technical communication research. Students tasked with researching technical communication topics that require understanding the topic extensively can use papers where researchers discuss their study and clarify ideas based on feedback—for example, the recent shift in how people communicate in a pandemic is the focus of many research, proceedings papers are spaces that researchers can give their most up-to-date findings, information coming directly from the researcher, rather than from a tertiary or non-scholarly source that summarizes the ideas.

Periodicals

Periodicals refer to multivolume, serial edits. This is simply documents published daily, weekly, or monthly to a specific publication. These can be newspapers or magazines. These sources are considered reputable because their standards of research depend on the reputation of the publishers. Articles are not subject to anonymous peer review; rather, the expertise of the author and (often) a review by the publication's editor are used to establish the quality of the research presented in articles. In technical communication, trade journals or trade magazines are periodicals and include publications such as *Intercom* magazine (published by the U.S.-based professional organization the Society for Technical Communication) and *tcWorld* (published by the German-based professional organization tekom).

Periodicals are numbered: volume is the number of years it has been published, and issue is the number of times it was published that year. Let us look at *Intercom*, the magazine published by the Society for Technical

Communication (STC), they use the volume and issue number. The first year it was published would be the first volume, and the issue refers to how many times it was published in that year. So the first issue of the magazine would be numbered 1. Consider then that the most current *Intercom* magazine (as of the writing of this chapter) is volume 67, issue 3. This means *Intercom* has been in publication sixty-seven years and published three issues for the year.

Periodicals have a similar publication process as books— they have an editor, copyeditor, and a style reviewer. And even though the periodical can only be accessed by having a subscription to the magazine or the organization to which it is affiliated, the audience is usually broader. For example, *Intercom* is meant for academic researchers and technical communicators who work in the industry (practitioners): both are authors and consumers of magazine articles as well.

Since the audience can be both researchers and practitioners, the articles share content that includes concepts applied to real world matters that makes it relevant for this broader audience. Consider, for example, the title of the most recent *Intercom* magazine: "To Boldly Go Where (Almost) No Technical Communicator Has Gone Before," where the author makes the connection between the function of technical communicators to pulp fiction writers. The information in its language and style is relevant to this broader audience.

As new researchers, you may find the language and style of trade magazines more accessible. These are great resources to begin your research. You should note, however, that unlike journals, due to the quick turnover to publish articles, there is no peer-review process to ensure accuracy of content in magazine articles. As researchers in higher education, a good practice is to cross check the content with other sources that have a more rigorous publication process.

Periodicals are a good source for researchers to start. Trade magazines present focused information on topics and they are usually based on personal experience or interviews with primary informants through interviews. Freshman researchers will find the language and applications relatable, and they act as bridges to understand the more difficult concepts and theories that academic researchers reference in scholarly books, journals, and conference proceedings papers where they share data that is collected in a systematic way. For this reason, periodicals like trade magazines are often a good resource when you wish to research topics to get the writer's perspective, learn about researchers' experiences, or see applications of academic research concepts in industry.

Websites/Online Documents

Of all the sources listed for research purposes, you may be most familiar with websites and other online documents. These sources are published by an organization or company and are accessible, either free of cost (advertisements usually pop-up or line the margins of the documents) or for a small subscription fee. Similar to periodicals, researchers consider websites and other online documents reputable sources, because their standards of research depend on the reputation of the publishers. Researchers often use websites and online documents to get background information to start their research: information about an individual, an organization, statistics, dates or events. A quick web search directly from the website of a person, place or thing you are investigating can be helpful. So, for example, if you are researching your college, you can go directly to its website and find information about when the college was established and names of various individuals in administrative positions.

It is important to know that the publication process of websites and online documents are not as rigorous as the other sources noted earlier and, depending on the author of the site, in-house processes can vary significantly. The various government departments, for example, will only publish information on their websites that has been confirmed accurate by regulatory bodies. Universities and other educational entities will have some type of vetting of materials published on their websites: these can be for style, legal parameters, and accuracy. Similarly, some business organizations also vet the information published on their websites for style, legal parameters, and accuracy. As a result, the credibility and reliability of these online sources is often a matter of the credibility and reputation of the organization that publishes them. For this reason, you should always identify and review the source (e.g., organization) that published online information to determine how credible, accurate, and objective their information might be.

The information on websites and other online documents are meant to be easily accessible information to the general public. As researchers, a good practice to assess the reliability and accuracy of content online is to look for an author, copyright date and/or date published, and information of who to contact about the content of the website. If one or more aspects of this information is missing, it is important to triangulate your information, this means you should cross-check what you found with information from other sources or contact the entity directly.

Websites and other online documents offer information on a wide range of topics, and they are easily accessible. For this reason, online documents are a good source to start your research, you can familiarize yourself with a topic and, if they cite sources, you can find more credible sources to look for more information. It is always an important strategy, however, to cross-reference the information gleaned from online sources with more credible sources like journals or books.

Blogs, Wikis, Social Media, and Other Self-Published Online Documents

Online publications can vary from self-published blogs by individuals providing music critiques and personal journaling to academic researchers and industry specialists writing on topics that vary from teaching strategies for technical communication online courses to initial findings about ongoing research. You could find specialized information by technical communication specialists on wiki pages describing theoretical concepts, and YouTube videos sharing complex "how-to's". The audience is generalists, varying from content specialists looking for quick indexical information to freshman researchers needing content.

As you may have noted with books, journals, and conference proceedings, due to the restricting publication process and subscription or other costs, both the writers and audience can be limited. Self-publishing your content online can cut those restrictions and open access to a more general audience to your content. Therefore, platforms like Wikis, YouTube, and social media (in general) are popular with audiences and authors who want to get their content out quickly and unrestrictedly.

Assessing these sources can be tricky. In some cases, the authors are well documented researchers using platforms other than the ones academics or researchers use to publish important information. For example, the astrophysicist Neil de Grasse Tyson often uses X (formerly known as Twitter) to share scientific information. An important strategy researchers use is to identify the authors of the information to determine the source's credibility and then triangulating that information with credible sources (sources written by content specialists) to determine accuracy. Again, if information is missing about the author, publisher or publication date, the source may be unreliable or inaccurate. An important duty of the researcher is to avoid using unreliable information in your research.

Blogs, wikis, social media, and other self-published online documents are another good source researchers can use. They can be used as a good database to judge audience reactions to certain topics or events or to directly contact researchers. Researchers even use these sites to understand communication strategies and the rates at how messages disperse to local and global communities. For these reasons, blogs, wikis, and other self-published online documents are a good source to familiarize yourself with a topic. It is important, however, to cross-reference the information with more credible sources like journals or books.

Standards Sources Should Meet

We have looked at the different types of sources you may come across when initiating your research. You can use this information to focus your research or to steer away from certain sources. Choosing reliable sources for your research is important because your readers depend on you for accurate information, which raises your ethos—your credibility as a writer and researcher.

At this point you may have a research topic, some keywords or phrases, and, possibly, a research question to start your research. Once you have identified sufficient sources for your research, there are a few things to keep in mind as you hone into these sources. Up to now, you may have done a great job identifying what seems like reliable sources. But often your technical communication research has to be kept within certain parameters. And you won't be able to determine if the sources so far are within the parameters unless you understand what those guidelines/parameters are. Here is a list to keep in mind. Are the sources you have chosen—

- Accurate
- Timely
- Relevant
- Focused
- Rigorously documented?

You will notice that assessing a source is a process. First, you selected various sources, then evaluated the types of sources for reliability, and, next, you will determine whether the source meets other academic guidelines.

Accuracy

As a new researcher, you will have to determine whether the information of a source is correct. You have an important tool in your toolkit to help determine accuracy—triangulating information. We noted triangulation earlier in determining whether you want to use a source. It includes looking at other sources to see if the information one source shares is similar to or contradicts another source. As you review the various sources chosen, start making annotations; this will help you remember what is being said from one source to the next.

If you recall, the publication process for some sources does not include a review for content accuracy. It is then up to the author to make sure they report accurate information. And even with sources that include the "anonymous-review" process in some cases, articles have been found with inaccurate information. Take for instance, the researcher Edward T. Hall who developed the concept of intercultural communication: high and low context cultures and listed descriptions about what makes one culture high context and others low. His research methods used to gather evidence were later discredited (Kittler et al.), and, so, some of his ideas were found inaccurate. Two ways of overcoming including inaccurate research are to:

1. Review what other researchers are saying about someone's work.
2. Use timely sources.

This last point is discussed next and is the second parameter to consider as you research.

Timely

Determining whether the information in a source is timely, so, not outdated, is a relatively easy process. It involves identifying the publication date of the source. You may be familiar with the different types of paper formatting: MLA, APA, and the like. Whichever style of formatting is required or preferred, one piece of information you will be asked to submit is the publication date. As noted earlier in assessing websites and other web-based documents, a date is an important piece of information to determine the reliability of a source.

In books, a page is dedicated to publication information: there you will find publisher information, publication date, cataloguing information, and ISBN numbers. Sometimes the month, date, and year are provided, other times only the year. For journals, conference proceeding papers, and periodicals, the volume, issue number, and date are provided at the top of the articles or on the introductions/cover pages of the sources themselves. For some web-documents, the date of publication can be found at the top of the document or at the close. If it is not evident in these areas, students can use the copyright date usually found in the footer of the document. Social media sources usually carry a time and date stamp.

Your research assignment would usually limit you to a specific time frame that your sources should be published. Many give a ten-year timeframe, so research completed in 2021 should use sources published by or after 2011. This aspect is helpful because, as noted earlier, newer research will help you determine the accuracy of past research. There is an exception though. Foundational research in a particular field can be outside of the ten-year limitation. Foundational works are those works in a field of research flagged for their extraordinary or original concept/contribution regardless of time, such as, the works of Michel Foucault. While he died in 1984 and his works were published before or during the 1980s, Foucault's writings today retain the status of foundational works and are cited extensively to date.

Relevant

Sources you have chosen may have titles that seem closely tied to your research, but the information of the source may not pertain to your subject matter and so is not relevant to your research. For example, while conducting research on mobile learning, I kept finding works about multimedia technology. At first glance, the information seemed pertinent, but upon further investigation the sources were not relevant to my research. Very often similar-sounding ideas may emerge in your search. A quick read of the source's abstract will give you an idea whether the document's topic truly informs your research. Be sure to assess titles of your sources as well as abstracts to help determine whether a source is actually relevant.

Focused

Once you have determined the source is relevant, technical communication research requires that your sources approach the topic in a focused fashion rather than in a superficial or generalized manner. Generalized information has its purpose, and often to initiate your research it is good to find general information to acquaint you with the topic. As the writing process advances, research that deeply engages with the *who, what, where, how,* and *why* of the subject will be indispensable giving credence to the importance of sources providing detailed, focused information. Sometimes this type of information may require more than one source. As you delve into your research, thereafter, it is important to assess your source for how focused it is in investigating and challenging the topic.

Rigorously Documented

As mentioned earlier, sometimes you need more than one source to complete the research. Previously accessed and assessed sources can direct you to other related sources. Sources you find must not only be focused on its content, but also the data should be well researched, so rigorously documented. This not only means the source gives credit to other researchers, but it helps readers find other sources and shows evidence that the information is well researched. As researchers, you appreciate the importance of learning how to cite your information, and how such citing helps guide other researchers in the right direction. One way of assessing a source for being rigorously documented, then, is to check the citation page. Look at the titles of the sources to match their relevance to the topic of the source you are currently evaluating.

Rhetorical Reading in Action

Let us connect the standards of reliable sources to reading rhetorically. As you begin to research, you will establish keywords and phrases. An example of these is in the memo assignment given earlier: "memos," "writing in professional contexts," "memo samples," and "types of memos." As your sources populate after inserting these keywords and/or phrases, you can quickly narrow your search limit to a ten-year timeframe within your library's interface. Once your list populates, before reading the sources, you can scan the abstracts to decide whether the sources are relevant to your research.

If done purposefully, you will have significantly narrowed down the number of sources for the research. Here, rhetorical reading can be utilized to appraise your sources. Identify the author; if no biography is in the source itself, do a quick web search. Find the author's background: does the author work in industry, have experience in technical communication? How is the writer's background relevant to the topic? You might consider this next step odd, but skim through the references list. Look at the number of sources and the title of the sources. Are the titles connected to the topic; do you see any sources that may be helpful to you? You may add additional sources to review. The reference list also helps you determine the rigor of the source's research.

Once you have determined the source to be relevant and reliable, focus on the argument the writer makes, what does the author use to make their argument (stories, statistics?), and how does the author connect to the reader? The answer to these questions helps you understand the information provided and to whom it is directed. You will not be able to remember all this information, so an important aspect of your research is annotating sources. The annotated bibliography is a handy tool to use, this is where you keep track of the sources and doubles as a way to cross-reference your sources—a final way to determine your sources' accuracy.

Assessing sources, as you may have noticed, is an important toolkit to develop your research practice. Again, the process altogether may seem painstaking, but using rhetorical reading strategies, understanding your sources, and being scrupulous in identifying and choosing sources puts you ahead and it comes easier with practice. Fortunately, you may have already mastered some of these skills, and the information here builds on these and facilitates your meeting the research rigor that college demands.

And there, you are on your way to becoming an ethical researcher.

Works Cited

Kittler, Markus G., et al. "Special Review Article: Beyond Culture or Beyond Control? Reviewing The Use of Hall's High-/Low-Context Concept." *International Journal of Cross Cultural Management*, vol. 11, no. 1, Apr. 2011, p. 63–82.

Miller-Cochran, Susan K., and Rochelle L. Rodrigo. *The Wadsworth Guide to Research*. Cengage Learning, 2014.

Society for Technical Communication. "Intercom: The Magazine of STC." *Society for Technical Communication*, May/June 2020, https://www.stc.org/intercom/2020/08/minimalism-revisited-copy/.

Teaching Resources

Overview and Teaching Strategies

Freshman students are often first introduced to both research and to technical communication as a field of study in their freshman courses. This chapter offers some tips to help them understand an important aspect of research: assessing sources they will encounter. Even students who have research background and technical communication experience will probably be new to the vast data resources available at higher education institutions. Without knowledge of how to use these resources, students may fall back on what they know which will not meet the requirements of research assignments for college courses or technical communication research assignments. Understanding how to find and assess sources, therefore, is pertinent for research.

Additionally, technical communication courses often provide the most relevant and practical ways that students get to practice the application of research concepts. For the most part, technical communication courses offer real world scenarios that students often find engaging and rewarding. The assignments offered in technical communication courses, therefore, easily lend to identifying concepts from the chapter and using these ideas allows for research practice that freshman students can build on for the rest of their academic careers.

Discussion Questions

To help students apply the concepts from the reading, consider having them discuss the following questions:

1. Which of the listed types of sources are you most and least familiar with? Share why you feel you may not be familiar with the source or else how you became acquainted with it.
2. Which of the five standards (accurate, timely, relevant, focused, rigorously documented) do you feel would be the hardest to determine from a source? Why?
3. List at least three ways that rhetorical reading will help you with assessing sources for technical communication research.
4. Consider a research paper you have pending. Explain the research process you plan on using to make sure you have reliable and credible sources.

3 Last to Be Written, First to Be Read: Writing Memos, Abstracts and Executive Summaries

K. Alex Ilyasova

So, you've had writing assignments before, probably ones where you had to compare and contrast something or a few things, or where you wrote a story/narrative.[1] And possibly a few where you've even had to write a research paper on a specific topic. You've maybe even had to write an argumentative paper at some point on some issue or topic as well. But now you're in a business writing class or a technical writing class, and things feel different. You now have to write a report. Not only do you have to write a report, but you also have to write these other parts of what seem to be pieces that go alongside the report, such as memos, and even smaller pieces like summaries and abstracts, that you might never have had to do in previous writing assignments. Where do you start? How do you know you did them well? And why do you need to write all those other pieces anyways?

What follows is a series of practical guidelines that help you understand the *why* and the *how* for three specific pieces of writing that often accompany reports and other professional and technical writing assignments. These three pieces of writing are cover memos, executive summaries, and abstracts. As you read the guidelines, you'll come away understanding the goal, purpose or why of each piece of writing, the definition, and what aspects make each effective. Along the way, we'll apply some of these guidelines to actual examples. You'll also be able to get into the heads of other writers as they formulate their own approaches to how they write these specific pieces/assignments.

---

[1] This work is licensed under the Creative Commons Attribution-NonCommercial-NoDerivatives 4.0 International License (CC BY-NC-ND 4.0) and is subject to the Writing Spaces Terms of Use. To view a copy of this license, visit http://creativecommons.org/licenses/by-nc-nd/4.0/, email info@creativecommons.org, or send a letter to Creative Commons, PO Box 1866, Mountain View, CA 94042, USA. To view the Writing Spaces Terms of Use, visit http://writingspaces.org/terms-of-use.

Lastly, one of the things these types of writings have in common is that they are often the last thing you write after you complete a phase of a project or the writing of a report, but they are one of the first things to be read by your supervisor, client, or other decision maker. As a result, in order to write these well you need to know and often to have completed a project or phase, or the report itself before you write these well, concisely, and persuasively. As you read through the guidelines for each type of writing, it will make sense about why these are written last.

Guidelines for Writing Cover Memos

Cover memos are a specific kind of memo that provide information about projects, tasks, and/or decisions to be made. These differ from more general memos in that they either accompany a report or reference a stage of a project. With this in mind, let's begin by looking at writing cover memos. To do so, we'll focus on three factors:

- Why we write cover memos or the purpose.
- What questions to answer/think about before you start.
- What are the main parts to include.

The objective of this process is to help you articulate information about project updates or completion, ongoing and/or completed tasks, decisions from important meetings, and/or short proposals that either you are working on by yourself or with a team.

Why We Write Cover Memos

To understand why we write cover memos, let's look at an example of one in Figure 3.1 and examine the following three aspects:

- **The content:** what is the topic or what does the memo discuss?
- **Parts:** what are the main sections of a memo? See if you can identify the four main sections—header, introduction, body, and conclusion.
- **Format:** how the information is presented (e.g., font type and size, how is the information arranged, length, etc.).

Additionally, see if you can tell whether or not this memo was for inside or outside of an organization, and the levels of formality and importance conveyed in the memo. Lastly, think about when you would write such a memo before or after you have completed a report?

HA Engineering Company

MEMORANDUM

Date: November 11, 2003

To: Henriette Alexander, CEO

From: Katherine Schmidt, Lead Engineer

Subject: Report on the solar options for C.Hill Industries

I have enclosed our report on C.Hill Industries' plans to convert its Springs facility to solar energy.

Overall, as our report shows, it looks like solar energy generators will work, but we have made some recommendations for improving the design. Specifically, we recommend they consider a design that more efficiently makes use of panel placement to improve energy yield. We believe this change would allow them to run the refrigeration units "off grid" during nonpeak hours.

Please review the enclosed report as soon as possible. We would like to schedule a conference call with their CEO, Lena Mack, in the next couple of weeks to discuss the report.

Thank you for your time. If you have any questions or comments, please call me at ext. 3111 or email me at k.schmidt@HA.net.

Figure 3.1. Memos are basically the same in structure as emails.

The first thing you might notice with the content is that the memo is written to someone inside the company or organization. This memo lets the reader know that a report accompanies it, and what the report is about. What this means is that you should write this kind of cover memo *after* your report is done. Cover memos contain information about project updates or completion, ongoing and/or completed tasks, decisions from important meetings, and shorts proposals that either you are working on by yourself or with a team.

One question students have about memos is what the difference is between them and sending emails. As you might have noticed, the formatting is quite similar (both share similar basic features: TO, FROM, SUBJECT, and a MESSAGE). However, the spread-ability or reach are the main differences between memos and emails. Emails can be sent instantaneously to many people by one person, memos have less reach or spread since they need to be printed on paper and delivered to one person or a few individuals. Usually, the reach or spread is determined by whether the email or memo is for someone outside or inside your organization. In today's workplace, emails have replaced memos as the quickest way to communicate *informal* messages, particularly to people inside your organization. For people outside the organization, emails are more often expected to have a higher level of formality. In other words, the differences between emails and memos are dependent on audience and purpose.

Memos, on the other hand, are typically more formal, taken more seriously, and as a result tend to be less likely to go unread or dismissed when they arrive on someone's desk. Receiving a paper memo signals that the message is too important or too proprietary to be sent via email. For example, you should probably write a memo if you include financial data, product research and development details, marketing strategies, basically any information that a company owns and/or does not want to share publicly.

Despite these differences, memos and emails share many of the same basic formatting features, and content. As you can see in the previous example, the main format features of a memo include:

- Header
- Introduction
- Body
- Conclusion.

Each section is key, for readers rely on such information to understand what is being asked of them and how urgent the information or request is in taking the next step.

Before You Start to Write

Because of the reach, formality, and importance associated with memos, when you begin writing the various sections, you should take some time to consider your readers and how they will use this information. Because memos are meant to quickly convey information, attention should be paid to include only relevant information. In other words, doing some planning and research will help ensure that you come off as a professional and that you include only need-to-know information.

Here are the 5Ws and How Questions to work through before you get started:

- Who is your reader? (Some examples include your boss/supervisor, a co-worker/team member, or a client.)
- Why are you writing to them? (Some reasons include because another phase of a project is complete, you've encountered a problem/opportunity, or you need something to continue.)
- What is your main point or what do you want your reader to do with this information? (Some examples, you are providing an update and no action is needed, you encountered a problem/opportunity and need to meet to discuss next steps, or you need something and are asking for approval.)
- Where and when will my memo be read? (In other words, what is the context in which your memo is read—for example, the physical, economic, political, and/or ethical factors that might influence how your reader interprets and responds to your memo? Or more specifically, what are the concerns of your reader as they read your memo?)
- How will the reader use this document in the future? (For example, will your memo be shared with others? Will it be filed/kept, and if so, by who/where? In other words, memos might turn up in unexpected/unintended places and so thinking about the public aspect of your memo and how it represents you and your work can help you make decisions about what you include, and how you state your message.)

Answering these questions is an important pre-writing step because it will help you be clear on your purpose, audience, and main point(s), and provide you with the basic outline before you start writing. This planning is not difficult, and most students tend to vaguely know the answers as they write, but since going through these questions will typically take only a handful of minutes, answering them before you start is useful in helping you be explicit and clear before you start about who you're writing to, why you're writing, and what your readers need to know.

Once you have answered these questions, you might consider having these answers handy and referring back to them to make sure you are staying on track as you write your memo.

Writing the Memo

Now that you have a clearer idea about who you're writing to and why you're writing, it's time to organize and draft your memo. As with most writing tasks, some memos will require more time and care to develop than others. However, as you get more familiar with this type of writing and gain more practice, writing memos will become easier and just another part of the flow of your workday. Part of what will help you get better at this writing task is understanding that some parts of the memo make predictable moves regardless of your purpose. Specifically,

- the *header* features the DATE, TO, FROM, and SUBJECT;
- the *introduction* introduces and sets up the topic of the memo; and
- the *conclusion* repeats make points for the reader.

Keeping this in mind and recalling what these moves include will allow you to spend more time on what you need to develop clearly and concisely—the body of the memo. Finally, memos are often one-two pages in length so that they can be read quickly, with only the most pertinent information included.

Header

Let's first take a look at the header and introduction (see Figure 3.2) and identify the predictable moves in each. In the header, there are two moves to get right. First, getting the features of the header correct, and second, having a clear and concise subject line (see Figure 3.2). The header must have the DATE, TO, FROM, and SUBJECT, in that order, and ideally, formatted so that the information in each line is aligned the same. Additionally, for memos, your initials must appear at the end of your name in the FROM line. (Side note: unlike letters, memos are NOT signed at the bottom. Instead, memos are initialed at the top, next to the name of the person who wrote the memo).

Lastly, with the SUBJECT line, you should offer a descriptive and specific phrase that describes the content of the memo. Simply writing, "Project" or "Update" is too vague. None of us like reading an email with a non-descript or boring title, and we likely skip over it since it might seem unimportant. Most readers look to the subject line first to determine if they want to read the memo and/or when to read the memo depending on the SUBJECT line. Giving more specific information, such as "Update on Project Springs" or "Accident at District #4" helps a reader make that decision.

INTRODUCTION

In the introduction (see Figure 3.2, first paragraph) you should make at least three of the moves listed here, depending on your reader and your purpose:

1. Identify the subject of the memo. (What the memo is about.)
2. State the purpose. (Why you are writing the memo and why the reader is reading it.)
3. State your main point. (What are the key ideas you want readers to remember in the memo?)
4. Offer some background information, depending on how much your reader needs to or already knows. (Provide them with the information needed to understand each main point in relation to the overall subject.)
5. Stress the importance of the subject, depending on how critical it is to act on your message/point. (Explain why the reader needs to know this information and what the reader needs to do with this information after reading the memo.)

Your *subject* and *purpose* should be stated within the first two to three sentences of your introduction in order to reduce how long it takes for your reader to know what the memo is about. Tell your reader, up front, what you are writing about. Do not assume that your reader will know just from the SUBJECT line or because they are expecting a memo from you. Followed by your *subject* and *purpose* is your *main point*, or the action you want your reader to take. You need to have all three of these moves in the first paragraph or introduction in the memo, see Figure 3.2.

As you might have noticed, the introduction (first paragraph) of the memo in Figure 3.2 also includes some background information, very brief, one sentence, "Last week, the Director's Board..." about where this idea came from. The writer also stresses the urgency or importance in addressing this concern "as soon as possible". In some cases, providing the background makes receiving the memo less confusing, and stressing the importance is warranted given the situation. Both of these moves are ones you will need to determine depending on the situation, however, moves 1, 2, and 3 are always required.

BODY: THE DETAILS

The body of the memo (see Figure 3.2) is where you provide your readers with the information they need to make a decision or take action. The body of the memo is often the largest part, and often takes up multiple paragraphs. This is where you develop your points by providing facts, examples, data, and/or reasoning for the decision or action you want taken. Referring back to the answers to the 5W and How Questions about why are you writing, what is your main point and what do you want your reader to do with this information at this point? Once you've done that, what information do you need to provide your reader for you to achieve your purpose?

As you begin writing, divide the *subject* of the memo into two to five major topics or points you want to discuss further with your reader. For example, in Figure 3.2 there are two main topics discussed—the current situation and a draft of the plan to discuss further. Each one is followed with some reasoning, facts, data or examples. In Figure 3.2, the current situation is stated directly and right away, along with the reasoning, (current situation stated clearly) "Despite our current situation of no one testing positive during the pandemic, strict mask requirements, and sanitizing procedures, (the reasoning starts here) the possibility still exists that the COVID-19 virus could be contracted by our employees and spread among our factory workers." Immediately next is a fact, from a trustworthy source that helps supports the reasoning, "The World Health Organization reported in February 2020 that asymptomatic individuals can spread the virus. In these cases, one asymptomatic individual can infect our factory floor employees pretty quickly."

Following this is the start of the next topic: draft of a plan. Your reader will expect clear reasoning and details in this section as well. For example, the section begins with the topics of this section and clear reasoning, "To help make sure everyone stays safe, informed and adhering to safety orders (the reasoning), we recommend the development of a contact-tracing plan (the topic of the section) for our company and workers." Lastly, the details or actions to take are also provided, "1. Develop a decision tree. . . . 2. Design an alert system. . . . 3. Connect with local health authorities. . . .".

When you include these elements—facts, examples, data, reasoning, actions—in a well-organized and clear way, you help your reader understand what you're asking of them and why.

Take your time with this section since it will provide the needed details to help your reader take action.

Conclusion

Finally, you're at the conclusion. The conclusion should be short, and like the rest of the memo, to the point (see Figure 3.2). Typically, you are looking at no more than three sentences, and nothing essential should be in the conclusion that hasn't already been stated in either the introduction or the body. As mentioned earlier, the conclusion is another place that has specific moves you can do quickly once you know them. These moves are:

1. Thank your reader: tell them you appreciate their attention.
2. Restate your main point: remind your reader of what you want them to do.
3. What's next: end with next steps and looking forward in some way.

Each of these moves is important because they meet the expectations of the reader. These expectations include indicating that the memo is concluding ("Thank you for your attention to this important matter," reminding the reader of the topic or urgency of the matter "I believe a contact-tracing plan should be developed as soon as possible," and providing contact information and a clear request for what is the next step, "To get started, I would like to schedule an appointment . . . you or your assistant can reach me at ext.19 . . . .".

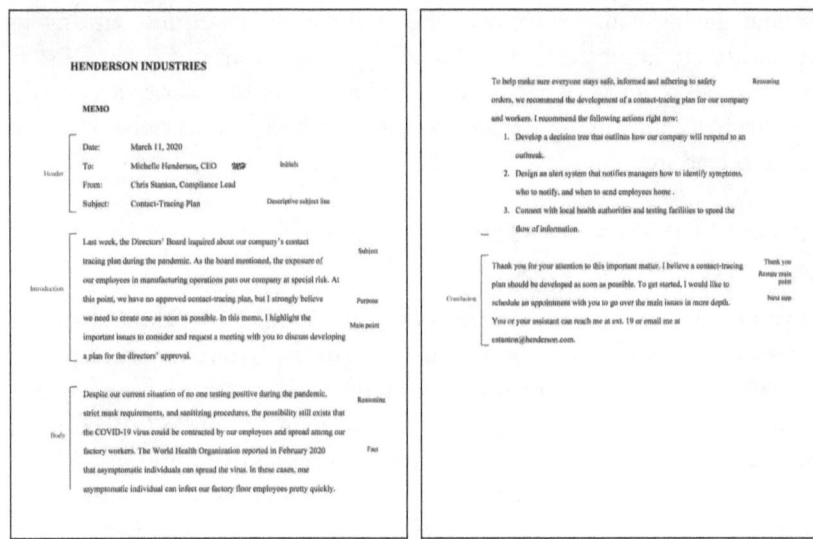

Figure 3.2. This memo shows the main parts and moves of the correspondence. The header indicates the subject of the memo; the introduction sets up the subject, purpose, and main point; the body provides the details; and the conclusion restates the main point and/or provides next steps.

Guidelines for Writing Abstracts and Executive Summaries

Next, let's review the process of writing abstracts and executive summaries. To examine these kinds of writing, we'll focus on three factors:

- Why we write abstracts or executive summaries,

- What to do before you write your abstract or executive summary, and
- What are the parts to an abstract or executive summary.

The objective of this process is to learn how to provide a brief and clear overview of an entire report without having your reader read the report itself.

Why We Write Abstracts and Executive Summaries

One of the writing assignments you will likely encounter in technical communication is report writing. And at some point, you will be tasked to write the front matter for a report, which will include either an abstract or an executive summary. Both the abstract and the executive summary are designed to provide a brief and clear overview of the entire report without having your reader read the report itself. Like most writing in technical communication, the goal is to provide clear and concise information quickly. That's what abstracts and executive summaries need to do well when accompanying a report—to inform readers about what they are about to read in the report.

Before You Start: Abstracts Versus Executive Summaries

As you might have already figured out, you can only write these abstracts and executive summaries once you have finished your report. Because both the abstracts and executive summaries reflect the complete content of your report, you must be done with your report in order to accurately reflect its content in either the abstract or the executive summary. Moreover, as you have also figured out by now, these pieces of writing will be the first things your readers will read in order to decide 1) if they want to read the rest of the report and/or 2) if they can make a decision based on the information in just the abstract or the executive summary. In other words, take your time drafting these as it will determine whether your reader proceeds with reading your report. Even though they are short or shorter than the report itself, abstract/executive summaries are the first impression readers will have about you, your work and professionalism, so they are quite crucial.

Table 3.1 provides some quick differences to keep in mind between abstracts and executive summaries.

Finally, before you get started drafting either an abstract or executive summary it is a good idea to do the following four steps as outlined by Purdue Online Writing Lab (see https://tinyurl.com/5n8za6bt).

Table 3.1. Abstracts Versus Executive Summaries

	Abstracts	Executive Summaries
Length	Typically about one robust paragraph.	Often about one page maximum.
Types	Two types: informative and descriptive. The descriptive does NOT include results, conclusions, or recommendations.	Only one type.
Order	Follows the organization of the report.	Does NOT follow the organization of the report.
Phrasing	Uses the phrasing in the report	Does NOT use the exact phrasing of the report.

Reread your report with the purpose of abstracting in mind. Look specifically for these main parts: purpose, methods, scope, results, conclusions, and recommendations. You might even want to mark your report so you can find those places once you start writing.

After you have finished rereading your report, write a rough draft without looking back at your report. Do not merely copy key sentences from your report. You will put in too much or too little information. Also, do not summarize information in a new way—no new information goes into either the abstract or executive summary.

You will likely find some weakness, wordiness, and errors. Revise your rough draft to:

- Correct weaknesses in organization and coherence,
- Drop superfluous information,
- Add important information originally left out,
- Eliminate wordiness, and
- Correct errors in grammar and mechanics.
- Carefully proofread your final copy.

Writing the Abstract or Executive Summary (Purdue OWL Staff)

Abstracts

When writing an abstract, draw on the key sentences directly from your report. Key sentences would be the ones that have a purpose statement, and then your main points. From there you should pull one or two key sentences from each major section of your report. This is where you might find key sentences that speak to your methodology, the results and/or discussion, and the conclusions or recommendations. Here is what this might look in terms of organization:

- Purpose statement (one sentence)
- Main point (one sentence)
- Methodology/scope (one or two sentences)
- Results and/or discussion (one or two sentences, include if you are writing a informative abstract)
- Recommendations/conclusions (one or two sentences, include if you are writing an informative abstract)

Remember that you should be using the original phrasing in the report as much as possible, modifying the sentences as needed to make it readable to your audience. At this point you might be wondering when to write an informative versus a descriptive abstract. When in doubt always ask your instructor. However, if you're allowed to decide here are two things to help, both require you understand your reader and context.

Writing a descriptive abstract or not including the results and recommendations/conclusions may entice your reader to look beyond the abstract and either read the report or skip to those sections. Writing an informative abstract or including results and recommendations/conclusions helps busy readers (such as CEOs, managers, and other decision makers pressed for time) to not have to read your report to get the full picture. Whichever one you chose, chose it consciously, knowing why and what result you are hoping for. Here are some discussion questions to help you decide:

- Who is your reader?
- How quickly do they need to know this information?
- Does your reader need to make a decision based on this information?
- Does your reader need to read the report completely and why?

Figure 3.3 provides an example of a scientific abstract. It contains all of the elements mentioned earlier for an effective abstract: purpose, main points, methodology, results, and conclusions/recommendations.

Executive Summaries

When writing your executive summary, you need to keep in mind two main aspects of executive summaries that differ from writing abstract. First, you should paraphrase the main sections and main points of your report versus trying to use the same phrasing from the report as you would in an abstract.

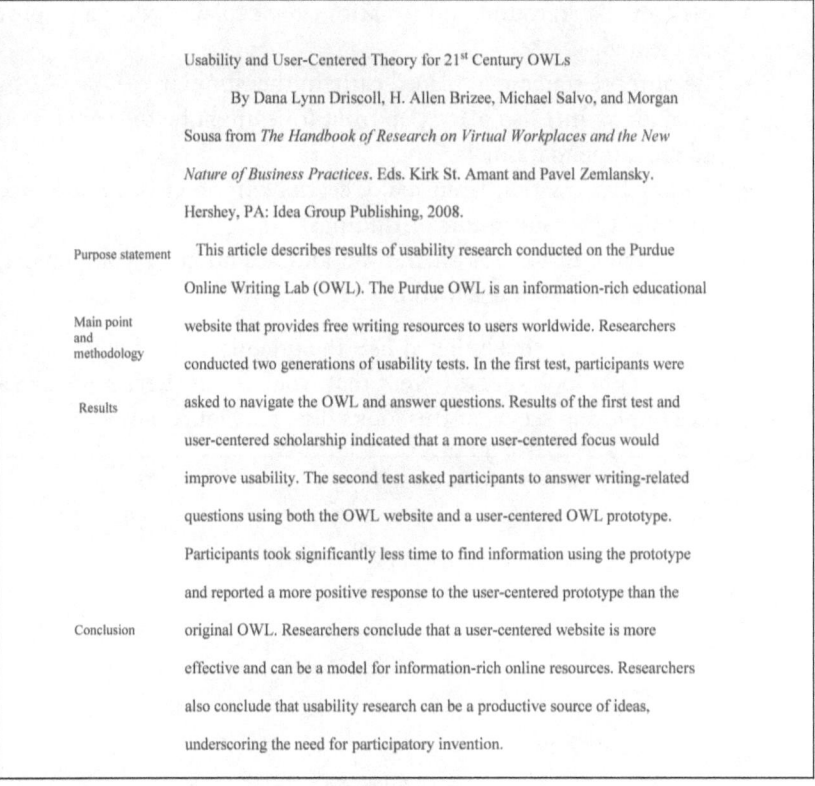

Figure 3.3. Example of a Scientific Abstract. (Based on https://owl.purdue.edu/owl/research_and_citation/using_research/writing_scientific_ abstracts_presentation.html)

Second, you should be thinking about the best way to organize your summary. This situation often means *not* following the organization of your report. Here are a few discussion questions to help you think through how to organize your summary:

- What does my reader need to know?
- What are the main points I need to highlight in order to inform my reader?
- What is an effective order that I organize these key points and why?

Each of these questions is important to answer because the answers will help you stay focused and well-organized.

Now that you've considered the needs of the reader and the most effective way to organize your executive summary, here are the items you should include:

- Relevant background information to explain why a report was created.
- The purpose statement, placed early in the summary
- The main point, also placed up front (this might be the conclusion and recommendation)
- The key information from major sections in order of importance (this might be your results or findings)
- Thank your reader for their time and attention to your report, along with a way to reach you.

As you can see from this bulleted list, the organization of your executive summary might look very different than your report. Let's now take a look at an example and see what this looks like (see Figure 3.4).

Executive Summary

Relevant background information	This report was written in response to a challenge in our Sustainability Student Group from Linda Kolton, Director of Sustainability at Colorado College. She asked us to
Purpose statement	develop options for converting one campus building to renewable energy sources. In this report, we discuss the possibility of converting Osborne Hall's heating system to solar.
Main point	We conclude that heating Osborne Hall with solar sources would require a combination
Key information organized in order of importance	of direct gain (skylights) and solar panels. The combination of these two solar technologies would ensure adequate heating for almost all the building's heating needs. A backup heater could be retained for sustained cold spells.
Key information organized in order of importance	To develop the information in this report, we followed a five-step research plan: 1) develop an evaluation criteria, 2) gather information on solar heating from industry already implementing such practices, 3) analyze the heating needs and heating costs of Osborne Hall, 4) conduct a literature review on solar energy in the past 10 years, and 5) analyze models of passive and active solar heating that would be appropriate for this building.
Key information organized in order of importance	The results of our research show that solar heating for Osborne Hall is possible, even for such an older building on campus. Our results also show that Osborne Hall can be a model for developing solar heating systems around campus, because it is one of the more difficult buildings to convert. We conclude with a cost analysis that indicates that solar heating would save the campus money in the long run. In the case of Osborne Hall, solar remodeling would pay for itself in 5-6 years.
Thank you and next step	We appreciate your taking time to read this report. If you have any questions or would like to meet with us, please call Diane Reese at 207-555-0001.

Figure 3.4. Example of an Executive Survey

The goal of an executive summary is that it is written in such a way that it can stand alone, apart from the rest of the report. And, as we saw, CEOs and managers may not have time to read your full report but still need to know key takeaways to make a decision. Hence, taking some time to figure out how to write an effective executive summary is a crucial skill for you to have. Here, writing and clear organization is key to drafting an effective executive summary.

Conclusion

This chapter has presented three common writing tasks in technical communication: memos, abstract, and executive summaries. These often accompany report writing and reports on work from projects. These types of writing, although typically read first, are written after your work is complete. They provide an opportunity to review your work and provide the needed information in a clear, concise, and meaningful way. Taking the time to learn how to do these relatively short writing tasks well can help you keep the necessary people informed and help guide decisions quickly.

Works Cited

Purdue OWL Staff. *Purdue Online Writing Lab.* 3 Aug. 2020, https://owl.purdue.edu/owl/subject_specific_writing/professional_technical_writing/technical_reports_and_report_abstracts/index.html.

—. *Writing a scientific abstract.* 3 Aug.2020, https://owl.purdue.edu/owl/research_and_citation/using_research/writing_scientific_abstracts_presentation.html.

Teacher Resources

Overview and Teaching Strategies

These writing tasks are often drafted at one or more of three points during a larger project, report, or team task:

- When the large or whole project/report is complete,
- When a phase or stage of a project is completed, or
- When an expected need, development occurs.

This chapter can be taught when discussing any of these situations. By focusing on the purpose, audience, and main point(s), this chapter provides students with a mechanism for examining their projects/reports, and writing concise and useful memos, abstracts, and executive summaries.

When teaching these tasks, instructors need to make students aware that:

- Each of these writing tasks have conventions in format, tone, and organization that need to be met to allow them to be read quickly.
- Each of these writing tasks should be completed after the larger task/project/report is complete.

To this end, students need to learn the core idea that for them to write these writing assignments well they must know their projects/reports well. The framework in this chapter for each writing task provides students with questions that help them examine their projects/reports, and a mechanism for pulling that understanding into concise and organized piece of writing.

DISCUSSION QUESTIONS

To help students explore and develop the ideas discussed in this chapter, consider having them address—as individuals, in small teams, or as a class—the following questions:

1. By knowing who you are writing to, your reader/audience, how does that help you shape the information in each section of a memo? In each section of the executive summary? How does the reader's/audience's position in the organization, background information on the project inform what you include?
2. This chapter provides strategies for organizing each of the three writing assignments. What role does the organization of your main points—in all three types of writing—play in terms of the ability of reader to transition to reading the main report/project and not be confused? What are some examples of poorly organizing your main points?
3. What are your assumptions about concise versus short pieces of writing? Is concise writing always short? What are aspects of a piece of writing that make it concise? What is challenging about writing concisely?
4. It is often tempting, when you are done with a large project/report, to quickly draft the smaller writing tasks that are discussed in the chapter. What are some reasons to remember/recall about why devoting the time and effort to these writing tasks well is important?

Examining these items can help students better reflect upon their projects/reports and consider how to apply the ideas presented in each section to crafting well-organized and concise memos, abstracts, and executive summaries.

4 Drafting Technical Definitions and Descriptions

Quan Zhou

Technical definitions and descriptions exist all around us.[1] Many documents for work and for leisure come with terms, mechanisms, and processes that are unfamiliar to readers. Consumer products provide user documentation that defines technological terms; scientific articles describe mechanisms that form the natural world; government documents explain processes of voting procedures. Defining these terms and describing these mechanisms and processes are necessary for readers to understand information and solve problems.

For many of you, the university is an exciting place with a life of its own. But it is also a complex structure with terms that are often new to the first-time college student. For example, you are typically required to take many kinds of courses that range from general education courses, core courses in your major, electives, and so forth. You must understand these terms and types of courses in order to plan your study accordingly. You will also meet an array of people — professors, deans, advisors, etc. — and you need to know what roles they each play in your college life. Technical definitions help you understand the environment around you and navigate your college experience.

Furthermore, you will need to know crucial processes and procedures you must follow to finish your degree. For instance, many degrees require the completion of an internship. An internship is typically a short-term work experience that extends your study and prepares you for professional life. To complete an internship, you usually need to secure an opportunity, work with a supervisor, and fulfill a set of requirements by your employer and your college. If you are new to internships, it is good to consult with a technical description provided by your school to help you make informed plans.

---

1 This work is licensed under the Creative Commons Attribution-NonCommercial-NoDerivatives 4.0 International License (CC BY-NC-ND 4.0) and is subject to the Writing Spaces Terms of Use. To view a copy of this license, visit http://creativecommons.org/licenses/by-nc-nd/4.0/, email info@creativecommons.org, or send a letter to Creative Commons, PO Box 1866, Mountain View, CA 94042, USA. To view the Writing Spaces Terms of Use, visit http://writingspaces.org/terms-of-use.

In technical communication, technical definitions and descriptions are indispensable. They help readers understand concepts effectively, perform tasks safely, and make decisions wisely. Without technical definitions and descriptions, most technical documents would fail.

Technical Definitions

Technical documents often come with scientific, technical, business, and other terms of which their readers have little to no prior knowledge. Technical definitions describe and explain technical terms and concepts in text and visuals. A technical definition can not only explain what a term means, but also where it comes from, what it is made of, and what distinguishes it from similar terms. It helps readers understand and differentiate unfamiliar or specialized terms in order to read a technical document and perform tasks. To draft a technical definition, you must analyze your audience and context, gather subject-matter information, determine the appropriate type of definition, and manage content coherence of a document.

Analyzing Your Audience and Context

Technical documents are written for specific audiences who have specific characteristics. As you plan for a technical definition, consider who your readers are, what they know, and what they need the definition for. What needs to be defined depends on your audience. You may have noticed that, when I introduced internship earlier, I provided a brief definition. I did so because those of you in your first two years of college might not be familiar with this term. However, if I were writing for seniors only, I might not need to define it.

Consider, too, your readers' purposes. Do they need to gain a quick understanding of a term? Do they need a thorough understanding in order to make complex decisions? Answers to these questions help you decide if you want to provide a short definition or an extended one.

While some concepts can be explained in one sentence, many warrant extended definitions to fully make sense. As I discuss later in this chapter, writers use examples, analogy, visuals, and other ways to enrich technical definitions. Does an example provide an adequate explanation? Is an analogy appropriate and effective? What kind of visual is appropriate for your

Drafting Technical Definitions and Descriptions 63

readers' background and level of knowledge? Getting your audience and context right enables you to draft a technical definition appropriately.

Gather Subject-Matter Information

Whether you are an expert on the subject matter or possess some knowledge, it is wise to gather information about the terms you are defining. Definitions make a lasting impact in the perception of concepts and terms, thus must be authoritatively and effectively written. The following are some common ways you can go about gathering subject-matter information.

- Looking up a term in reference books: Many terms are defined in dictionaries, encyclopedias, and even popular sources. General reference books like the Webster's Dictionary, the Oxford English Dictionary, and the Encyclopedia Britannica are good places to start. If you work in a specialized field, consider specialized references as they cover domain-specific terms more comprehensively. Oxford's A Dictionary of Computer Science provides all-encompassing coverage of computing terms. WebMD.com covers thousands of health conditions and diseases. In addition, popular sources such as Wikipedia, wikiHow.com, and howstuffworks.com provide a large set of definitions, from everyday terms to uncommon ones.
- Synthesizing information from documents: Some definitions require you to gather and synthesize information from references, journals, Web articles, and other documents. Documents offer deeper knowledge and more comprehensive explanations than simple dictionary definitions. And because documents can treat a term differently for different audiences, they demonstrate many aspects and varying details of a term.
- Consulting with subject-matter experts: Many terms at work are best defined by subject-matter experts. These experts understand the context of a technical term and can offer a definition that best fits your goals. And if experts are not good at communicating their ideas, observe how they perform tasks and operate procedures. These first-hand experiences help you achieve greater accuracy and clarity.

Whether it's the help of reference books, pertinent documents, or subject-matter experts, the key to remember is that your technical definition must enable readers to use information effectively and safely. When in doubt, gather information on your terms and concepts.

Selecting the Appropriate Type of Definition

There are generally two categories of technical definitions: sentence definitions and extended definitions. Sentence definitions are short, often one-sentence, descriptions of a technical term. They are often the most typical definition for the everyday reader. Extended definitions are longer, more in-depth explanations.

Sentence Definition

A sentence definition is a brief and straightforward explanation of a term. It usually consists of the term being defined, the category it belongs to, and the characteristics that distinguish the term from other terms.

Consider the following definition of mortgage:

> A mortgage is a loan from a bank or other financial institution that helps a borrower purchase a home. (Taylor, "What Is A Mortgage?")

In this definition, "mortgage" is the term and "loan" is the category. The distinguishing characteristic is that mortgage loan helps a borrower purchase a home. This characteristic sets mortgage aside from other types of loans.

Selecting an appropriate category can be tricky. Category must provide concrete meaning and, at the same time, avoid merely repeating the term being defined. Take, for instance, the following example of a definition.

> A fixed-rate mortgage is a mortgage with an interest rate that stays the same. (adapted from Lewis, "Fixed-Rate Mortgages: What You Need to Know")

Here, the category mortgage merely repeats the term. It does not provide essential meaning that would improve readers' understanding of the term. This type of definition is considered circular definition, an explanation that uses the term being defined as the definition. Circular definitions appear as though the writer assumes that readers have prior knowledge in the term being defined.

How about the following alternative:

> A fixed-rate mortgage is a financial product with an interest rate that stays the same. (adapted from Lewis, "Fixed-Rate Mortgages: What You Need to Know")

This definition avoids being circular but isn't quite right yet. The category financial product is too broad to convey essential meaning.

Let's narrow down the category further.

> A fixed-rate mortgage is a home loan with an interest rate that stays the same. (adapted from Lewis, "Fixed-Rate Mortgages: What You Need to Know")

This version clearly categorizes a mortgage as a loan and then distinguishes mortgage loan from other loans.

Extended Definition

Unlike sentence definitions, extended definitions provide longer and more comprehensive explanations of a term. Here we discuss common strategies to extend a definition in various situations.

Extend by Example

Examples are an effective and powerful way to explain terms that are vague, abstract, complex, easily confused, or too specialized for the regular reader.

The following definition of text-to-speech technology is accurate, but to readers new to this technology, it is still abstract.

> Text-to-speech (TTS) is a type of assistive technology that reads digital text aloud. ("Text-To-Speech Technology: What It Is and How It Works")

For instance, where is text-to-speech technology used? Does it read text on your computer? Mobile phone? A few examples would do the trick.

> Text-to-speech technology can turn words and paragraphs on your computer, smart phone, tablets, and other digital devices into audio. (adapted from "Text-To-Speech Technology: What It Is and How It Works")

Choosing examples takes careful and thoughtful effort. Examples must represent the most typical ways a term can be fleshed out. Choose common, not rare examples. Examples must also be familiar to your readers. Here your knowledge and analysis of your readers come in handy.

Extend by Visual

You may have heard of the expression: A picture is worth a thousand words. Visuals can explain complex terms, spatial relationships, geographical distributions, or other concepts that are otherwise painstaking to explain in text. These days, readers are accustomed to infographics, animated visuals, and even interactive visuals. They expect a visual when it is needed.

Consider the following definition of the exploded view drawing, a type of visuals that is commonly used in technical instructions.

> An exploded view drawing is a picture or diagram that shows how an object is assembled.

Good luck with that definition. It's not that the writer did a poor job; however, defining this type of concept is best done with visuals. Figure 4.1 allows most readers to understand exploded view drawings immediately.

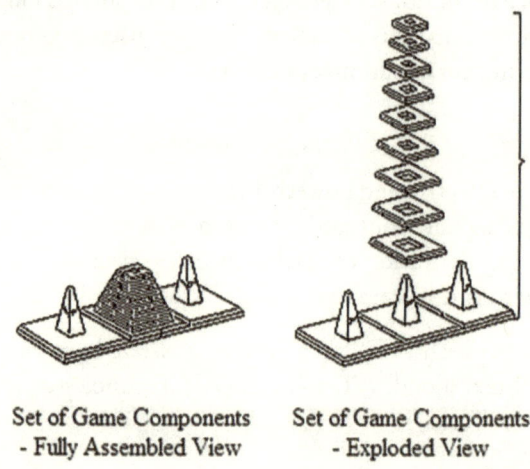

Set of Game Components
- Fully Assembled View

Set of Game Components
- Exploded View

Figure 4.1: An exploded view drawing (Source: United States Patent and Trademark Office)

When using visuals, remember that you must consider copyright issues. You should not simply use images from the Web, social media, or other authors without getting permission.

Extend by Partition

Many complex terms can be defined by partitioning their components. Partition elaborates on the categories, variations, and applications of a term. Remember that a term can often be partitioned in more than one way. Choose components that are major to the term and context. Figure 4.2 shows an extended definition of renewable energy by the U.S. Energy Information Administration.

This definition begins with a sentence definition of renewable energy and extends to the five forms (partitions) of renewable energy: Biomass, hydropower, geothermal, wind, and solar.

Drafting Technical Definitions and Descriptions 67

> **What is renewable energy?**
> Renewable energy is energy from sources that are naturally replenishing but flow-limited; renewable resources are virtually inexhaustible in duration but limited in the amount of energy that is available per unit of time.
>
> The major types of renewable energy sources are
> - Biomass
>   - Wood and wood waste
>   - Municipal solid waste
>   - Landfill gas and biogas
>   - Ethanol
>   - Biodiesel
> - Hydropower
> - Geothermal
> - Wind
> - Solar

Figure 4.2: Renewable energy definition using partitions (Source: U.S. Energy Information Administration)

Extend by Comparison and Contrast

When the term being defined is similar to, commonly associated with, or easily confused with other terms, consider extending the definition through comparison and contrast. What makes the term distinctive? How does the term resemble other terms? How does it differ from other terms? Comparison and contrast help readers distinguish the term being defined from other terms. They help readers gain a deeper understanding of the term.

Take a look at an excerpt of a definition of laminate floors in Figure 4.3. When homeowners consider flooring options, they often must decide among several common flooring materials, including laminate, vinyl, solid hardwood, engineered wood, and natural stone. Homeowners might need to know how laminate floors resemble and differ from other flooring types with regard to their characteristics, quality, installation requirements, and price. Figure 4.3 shows an effective use of comparison and contrast in an extended definition of laminate floors.

Extend by Analogy

A distinctive way of defining a term is making analogies. Analogy links abstract or unfamiliar terms with concrete and familiar ones.

Consider the following definition of "keyboard":

> A computer keyboard is one of the primary input devices used with a computer. Similar to an electric typewriter, a keyboard is composed of buttons that create letters, numbers, and symbols, as well as perform other functions ("Keyboard")

Laminate Floors Compared to Other Floor Coverings		
	Different From Laminate	Similar to Laminate
Vinyl Floor	Vinyl flooring is flexible, contains only vinyl product, and is 100-percent impervious to water. Vinyl flooring does not need to acclimate to a room prior to installation.	Vinyl is a close cousin of laminate. It is competitively priced, equally easy for do-it-yourselfers to install, and has a similar look.
Solid Hardwood	Hardwood is 100-percent solid wood. Laminate has no solid wood. Solid hardwood is thick and can be sanded and re-sanded many times. Laminate is thin and can never be sanded.	Solid hardwood and laminate flooring can look remarkably alike, especially from a distance. High-definition imaging techniques make some laminate flooring a dead-ringer for real hardwood.
Engineered Wood	Engineered wood has a plywood base topped with a veneer of 100-percent real wood. Laminate has no plywood and no natural real wood veneer top.	Both engineered wood and laminate have a base that made of manufactured wood. Both products can look remarkably similar, especially with the premium laminates.
Natural Stone	Laminate flooring contains no stone product. Stone is hard, solid, and thick. Laminate is flexible, breakable, and thin.	As with the hardwood-to-laminate comparison, higher-end laminate flooring can look very much like stone.

Figure 4.3: A definition of laminate floor based on comparison (What Are Laminate Floors?)

Here the writer thoughtfully uses the analogy of an electric typewriter for readers who are not computer savvy. Analogy is challenging. In this definition, for instance, the writer clearly assumes that more readers are familiar with electric typewriters than keyboards. This assumption might ring true these days. A few decades from now, it perhaps will not. Readers might not need a definition of a keyboard, nor might they be familiar with electric typewriters, except having seen one in museums. Appropriate analogy requires savvy understandings of the audience and context.

Extend by Etymology

Terms can be defined by etymology, or their origin. The origin of a term helps readers understand who invented the term and how the term evolved. The etymology strategy is particularly suitable for terms translated from another language or those that have gone through significant changes.

Consider this definition of emoji:

> An emoji (絵文字) which translates to "picture character" is an electronic pictograph (picture conveying a message) initially used in Japan, and now all over the world. Emojis were first invented by Shigetaka Kurita and became more mainstream when they were introduced on the iPhone in 2011. ("Emoji")

Emoji isn't native to the English language. This definition explains the origin of emoji and how it became popular in the English-speaking world. Consider etymology for terms whose origin holds the key to their meaning.

Extend by Negation

Sometimes, a term is defined not by what it is but what it is not. This strategy is considered negation. Much less common, negation can be used to define terms that are commonly misunderstood or confused with other terms.

Take, for example, the coronavirus pandemic that unfolded in late 2019. In the early days of the pandemic, there was widespread confusion between the novel coronavirus and the flu. The Center for Disease Control and Prevention provides the following definition of novel coronavirus, while stressing that COVID-19 isn't the same as the common cold.

> A novel coronavirus is a new coronavirus that has not been previously identified. The virus causing coronavirus disease 2019 (COVID-19) is not the same as the coronaviruses that commonly circulate among humans and cause mild illness, like the common cold. ("About COVID-19")

Negation helps clarify concepts that the readers are easily confused with or hold widely misunderstood notions of. Use negation scarcely and only when it is necessary.

Managing Content Coherence

Definitions don't exist in isolation from the rest of a document. They must serve a document's coherence and enable readers to move through the document effectively.

Where does the reader need a definition of a term? Is the definition stand-alone, part of a text or a longer publication? What has been said before a definition is introduced? What needs to be said after? Does a definition bring in new terms that need to be defined? Is a definition essential to the understanding of the text that follows? Is a definition helpful to know but not essential?

A stand-alone article about solar energy might begin with a definition of solar energy. But an article on solar energy in a broader publication about government energy policy might not. Often, the writer chooses to contextualize a term, introduce its background, explain its relevance, and then lead the reader to the definition. Without context, a definition of a new term could seem out of nowhere. Contextualizing a term helps readers understand why the term is being defined and how it relates to other topics in a document.

If you are writing a longer document, consider what has been said before a term is introduced. Consider the following definition of autism by the CDC.

> Autism spectrum disorder (ASD) is a developmental disability that can cause significant social, communication and behavioral challenges. ("What is Autism Spectrum Disorder?")

Has developmental disability been defined or explained prior to this definition in the document? If it has, readers might not need further definition. On the CDC website, however, this definition is in a stand-alone article about autism, so the CDC further provides the following definition of developmental disability.

> Developmental disabilities are a group of conditions due to an impairment in physical, learning, language, or behavior areas. ("Facts about Developmental Disabilities")

When a technical definition of one term brings up other terms that need to be defined, consider using a parenthetical definition to any term in need of further definition. That is, readers can quickly understand a term that would otherwise get in the way of understanding the primary term being defined.

The following is the result of a parenthetical definition applied to "developmental disability":

> Autism spectrum disorder (ASD) is a developmental disability (conditions due to an impairment in physical, learning, language, or behavior areas) that can cause significant social, communication

and behavioral challenges. (adapted from "What is Autism Spectrum Disorder?")

Consider, too, what needs to be said after defining a term. When the definition of autism introduces social, communication, and behavioral challenges, readers expect to know more about each of these challenges.

Where a definition appears in a document also matters. Important definitions, those that readers must have, are usually part of a document's main body of text. These important definitions often serve as opening sentences or thesis statements and are often given a distinctive formatting to visually distinguish them from the rest of the text. Some definitions appear on the page margin. They are supplemental to understanding the text. Still others appear in a footnote or glossary at the end of the document. These less important definitions serve as references.

To put it altogether, you should determine where a definition is needed, how important it is, and how they appear in your document.

Technical Descriptions

Technical description explains a technical object, mechanism, or process using text and visuals. Generally, technical descriptions explain how an object functions, how a mechanism works, or how a process takes place. An e-commerce website explains a new microwave oven for customers; a trade magazine describes the mechanism of carbon emissions for professionals; a government brochure describes the process to apply for a passport for the public. Technical descriptions not only inform readers about technical concepts but also educate the public about important issues. To draft a technical description, you must analyze your audience and context, gather subject-matter information, select the appropriate type of description, and develop effective details.

Analyzing Your Audience and Context

The first step in drafting a technical description is analyzing your audience and context. Unlike definitions, technical descriptions are longer, more thorough explanations. They give the writer greater flexibility with regard to what and how to draft, but they also require excellent understanding of your audience, purpose, and context. The audience and context of technical descriptions vary greatly.

Consider what your readers already know. Are they new to the technical mechanism or process you are describing? Do they have limited knowledge? Do they hold common misunderstandings? Suppose you are writing a description of COVID-19 vaccines for a trade magazine aimed at a general audience. To help this audience understand how these vaccines work, you might need to explain how the human immune system works. But if you are writing this description for medical professionals, many of whom are familiar with the human immune system, you might not need as much foundational scientific knowledge. Instead, you might consider topics such as administering vaccines and screening for precautions.

In addition, readers with different goals have their own expectations for technical descriptions. If readers are doing a casual read to get to know a concept, they might expect a description that is to the point and short enough that they can finish in a few minutes. If customers are doing thorough research on several products, they might expect detailed descriptions that guide their decision making. Yet other readers might expect to know all the details of a process well enough so that they can carry it out.

Furthermore, the context in which a technical description is needed determines how it is written. A description of the fire evacuation process is usually needed in an emergency situation where readers must take quick action. Such a process description must be easy to read and visually salient. On the other hand, a description of how to take care of a plant is often kept by gardeners as a reference to consult with from time to time.

Gather Subject-Matter Information

Technical descriptions often require substantial support of pertinent information. Gathering such information can be crucial as it gives you a thorough understanding of a mechanism or process. You can do so by synthesizing documents and gathering information with subject-matter experts.

- Synthesizing information from documents: Many documents and sources provide existing descriptions of the mechanism or process you are describing. If you are describing a common concept, such as how mortgages work, you can find plenty of publications and online resources. Publications such as the "for dummies" book series cover a wide range of topics. Websites like howstuffworks.com and wikiHow.com cover an even greater range of topics. If you are describing a specialized concept, such as how Mask Aligners work, consider technical and scientific magazines and specialized journals.
- Consulting with subject-matter experts: No one is an expert on every mechanism or process. When writing technical descriptions, subject-matter experts are some of your best information sources.

To know how something works, shadow subject-matter experts, observe how they work, and take detailed notes. Technical descriptions aren't done in one shot. After you draft your description, you will need to check with subject-matter experts to revise it for technical accuracy.

You might think you already know everything about a technical mechanism or process. But describing it to others is a different feat. As you dive further into a topic, you might uncover blind spots, clear misunderstandings, or learn whole new knowledge. Be prepared and allow yourself enough time to gather helpful information.

Selecting the Appropriate Type of Description

In general, there are two types of technical descriptions: Mechanism descriptions and process descriptions. Some also consider instructions as a form of technical description. For the purpose of this book, we do not cover instructions in this chapter.

Mechanism Descriptions

A mechanism or object description explains how a mechanism or an object functions. Mechanism descriptions can explain how a machine works, how a biological mechanism takes place, or how an engineering system operates. They are suitable for but not limited to the following topics:

- Scientific mechanisms: such as how human metabolism works.
- Technical objects: such as how the sawmill works.
- Technical and engineering mechanisms: such as how the microwave oven works.
- Professional and business mechanisms: such as how compound interest works.

Excellent mechanism descriptions enable readers to understand how an overall mechanism works and reduce unnecessary confusion.

Process Descriptions

A process description explains how a process works or how to carry it out. Process descriptions can explain how a phenomenon occurs, how a product is manufactured, or how a task is accomplished step-by-step. They are suitable for but not limited to the following topics:

- Technical processes: such as software development processes.
- Business processes: such as how to apply for paid leave.

- Legal and political processes: such as how to register to vote.

Excellent process descriptions help readers understand the scale, length, and complexity in a process and prepare for it informatively.

Developing Effective Details

An excellent technical description requires an effective structure and appropriate details. While mechanism and process descriptions share a lot in common, they have clear distinctions.

Mechanism Descriptions

Begin your mechanism description with your purpose, context, and scope. What does this description accomplish? What will readers take away from it? How in-depth is it? If readers hope to gain a basic understanding of a mechanism in a few minutes, provide a short description with basic facts and practical information. If readers are conducting research on a mechanism, perhaps provide a thorough description with comprehensive facts, background, and applications of that mechanism.

Introduce the mechanism and describe its most relevant aspects. What is it? What does it do? What does it look like? How does it work? How does it compare to similar mechanisms? Depending on what you are describing, many common strategies can be employed. For instance, you can describe a mechanism by its components and how they work together. You can describe a mechanism with regard to its causes and effects, problems and solutions. Another strategy is to compare and contrast a mechanism with other mechanisms for their similarities, differences, or commonly held misunderstandings.

Conclude your description by summarizing the major aspects of the mechanism. Depending on the length of your description, your conclusion could be as short as a sentence or two. It can also provide a relatively finer level of detail that highlights the gist of the mechanism.

Table 4.1 is an example that describes how text-to-speech technology works. This mechanism description first provides a summary of the most important points. Then, the description begins by introducing and defining text-to-speech technology. It clearly reveals its focus on children and identifies parents as the intended audience. It then describes how text-to-speech technology works and how parents can adopt this technology for children. It concludes with a list of key takeaways. This description is entirely appropriate for the intended audience, with the right length, level of detail, and selection of topics.

Table 4.1. A Mechanism Description (*"Text-To-Speech Technology: What It Is and How It Works"*)

At a Glance • Text-to-speech (TTS) technology reads aloud digital text—the words on computers, smartphones and tablets. • TTS can help kids who struggle with reading. • There are TTS tools available for nearly every digital device.	This mechanism description begins with a summary of important points. Notice that the summary clearly identifies the intended audience
Text-to-speech (TTS) is a type of assistive technology that reads digital text aloud. It's sometimes called "read aloud" technology. With a click of a button or the touch of a finger, TTS can take words on a computer or other digital device and convert them into audio. TTS is very helpful for kids who struggle with reading. But it can also help kids with writing and editing, and even focusing.	The author defines text-to-speech technology. The author then elaborates on TTS and explains their intended audience and purpose.
How Text-to-Speech Works TTS works with nearly every personal digital device, including computers, smartphones and tablets. All kinds of text files can be read aloud, including Word and Pages documents. Even online web pages can be read aloud. The voice in TTS is computer-generated, and reading speed can usually be sped up or slowed down. Voice quality varies, but some voices sound human. There are even computer-generated voices that sound like children speaking. Many TTS tools highlight words as they are read aloud. This allows kids to see text and hear it at the same time. Some TTS tools also have a technology called optical character recognition (OCR). OCR allows TTS tools to read text aloud from images. For example, your child could take a photo of a street sign and have the words on the sign turned into audio.	The author describes the nuts and bolts of how TTS works. Notice how the author connects the children with this description. The author also defines another term, OCR.
How Text-to-Speech Can Help Your Child Print materials in the classroom—like books and handouts—can create obstacles for kids with reading issues. That's because some kids struggle with decoding and understanding printed words on the page. Using digital text with TTS helps remove these barriers.	Here the author begins to explain how TTS can help children. Notice that the heading confirms that parents are the intended audience.

Table 4.1. A Mechanism Description (continued)

How Text-to-Speech Can Help Your Child (continued)

And since TTS lets kids both see and hear text when reading, it creates a multisensory reading experience. Researchers have found that the combination of seeing and hearing text when reading:

- Improves word recognition
- Increases the ability to pay attention and remember information while reading
- Allows kids to focus on comprehension instead of sounding out words
- Increases kids' staying power for reading assignments
- Helps kids recognize and fix errors in their own writing

Like audiobooks, TTS won't slow down the development of your child's reading skills.

Types of Text-to-Speech Tools

Depending on the device your child uses, there are many different TTS tools:

- **Built-in text-to-speech:** Many devices have built-in TTS tools. This includes desktop and laptop computers, smartphones and digital tablets and Chrome. Your child can use this TTS without purchasing special apps or software.

- **Web-based tools:** Some websites have TTS tools on-site. For instance, you can turn on our website's "Reading Assist" tool, located in the lower left corner of your screen, to have this webpage read aloud. Also, kids with dyslexia may qualify for a free Bookshare account with digital books that can be read with TTS. (Bookshare is a program of Understood founding partner Benetech.) There are also free TTS tools available online.

- **Text-to-speech apps:** Kids can also download TTS apps on smartphones and digital tablets. These apps often have special features like text highlighting in different colors and OCR. Some examples include Voice Dream Reader, Claro ScanPen and Office Lens.

> Here the author elaborates on types of TTS tools and example programs. This section can guide parents to select the appropriate program.

Table 4.1. A Mechanism Description (continued)

Types of Text-to-Speech Tools (continued)

- **Chrome tools:** Chrome is a relatively new platform with several TTS tools. These include Read&Write for Google Chrome and Snap&Read Universal. You can use these tools on a Chromebook or any computer with the Chrome browser. See more Chrome tools to help with reading.

- **Text-to-speech software programs:** There are also several literacy software programs for desktop and laptop computers. In addition to other reading and writing tools, many of these programs have TTS. Examples include Kurzweil 3000, ClaroRead and Read&Write. Microsoft's Immersive Reader tool also has TTS. It can be found in programs like OneNote and Word. See more examples of software for kids with reading issues.

How Your Child Can Access Text-to-Speech at School

It's a good idea to start the conversation with your child's teacher if you think your child would benefit from TTS. If your child has an Individualized Education Program(IEP) or a 504 plan, your child has a right to the assistive technology she needs to learn. But even without an IEP or a 504 plan, a school may be willing to provide TTS if it can help your child.

You can also use TTS at home. Try one of the previously mentioned tools, or check out options for free audiobooks and digital TTS books. And learn more about assistive technology for reading.

Key Takeaways

- Text-to-speech (TTS) can provide a multisensory reading experience that combines seeing with hearing.
- Using TTS won't delay the development of your child's reading skills.
- Your child's school can provide TTS, but you can also try it at home.

Finally, the author concludes this description with a summary of important points. Notice that this summary differs from the one at the beginning. It focuses more on actions parents can take.

Process Descriptions

Begin your description by introducing the process, your purpose, and scope. Describe the overall process and provide necessary definitions. State your purpose and the scope of your description. Does your description provide a basic outline of a process for readers who have an interest? Is your description here to help readers carry out a process with all the details needed? Does your description cover only major steps of a process, or does it also cover less important and miscellaneous details?

Provide an overview of the steps of the process. The overview lays out the steps in chronological order. It can serve as a map of the upcoming content.

Walk your readers through the process step-by-step. For each step, define its purpose, explain its context, and provide appropriate details. Often a step consists of sub-steps. Provide a clear hierarchy of steps to guide readers through the process.

Conclude your description by signaling the completion of the process and summarizing the steps of the process.

Table 4.2 provides an example that describes the DNA replication process. The description first defines the DNA replication process, outlines its three steps, and introduces DNA structure as a prerequisite for understanding DNA replication. It then describes each of the three steps of the replication process chronologically, using visuals at key places. Lastly, it describes what happens when this replication process goes wrong.

Table 4.2. A Process Description (Dovey, *"DNA Replication"*)

DNA replication, also known as semi-conservative replication, is the process by which DNA is essentially doubled. It is an important process that takes place within the dividing cell.	**The author defines DNA replication and introduces the DNA replication process.**
In this article, we shall look briefly at the structure of DNA, at the precise steps involved in replicating DNA (initiation, elongation and termination), and the clinical consequences that can occur when this goes wrong.	
DNA Structure DNA is made up of millions of nucleotides. These are molecules composed of a deoxyribose sugar, with a phosphate and a base (or nucleobase) attached to it. These nucleotides are attached to each other in strands via phosphodiester bonds to form a 'sugar-phosphate backbone'.	**The author describes the DNA structure before explaining the replication process.**

Table 4.2. A Process Description (continued)

The bond formed is between the third carbon atom on the deoxyribose sugar of one nucleotide (henceforth known as the 3') and the fifth carbon atom of another sugar on the next nucleotide (known as the 5'). N.B: 3' is pronounced 'three prime' and 5' is pronounced 'five prime'. There are two strands running in opposite or antiparallel directions to each other. These are attached to each other throughout the length of the strand through the bases on each nucleotide. There are 4 different bases associated with DNA; Cytosine, Guanine, Adenine, and Thymine. In normal DNA strands, Cytosine binds to Guanine, and Adenine binds to Thymine. The two strands together form a double helix. 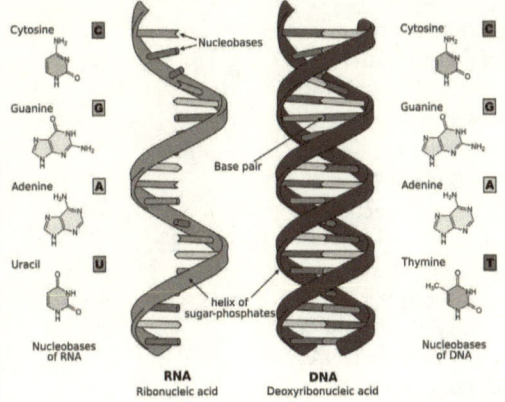 Fig 1.0. The Structure of RNA and DNA (By Difference_DNA_RNA-DE[CC BY-SA 3.0], via Wikimedia Commons)	 Visuals are used to explain the complex DNA structure.
Stages of DNA replication DNA replication can be thought of in three stages; Initiation, Elongation, Termination	The author lays out the linear process that consists of three stages.
Initiation DNA synthesis is initiated at particular points within the DNA strand known as 'origins', which are specific coding regions. These origins are targeted by initiator proteins, which go on to recruit more proteins that help aid the replication process, forming a replication complex around the DNA origin.	Using each stage name as a heading, the author provides details about the three stages.

Table 4.2. A Process Description (continued)

There are multiple origin sites, and when replication of DNA begins, these sites are referred to as replication forks. Within the replication complex is the enzyme DNA Helicase, which unwinds the double helix and exposes each of the two strands, so that they can be used as a template for replication. It does this by hydrolysing the ATP used to form the bonds between the nucleobases, therefore breaking the bond holding the two strands together.

DNA Primase is another enzyme that is important in DNA replication. It synthesises a small RNA primer, which acts as a 'kick-starter' for DNA Polymerase. DNA Polymerase is the enzyme that is ultimately responsible for the creation and expansion of the new strands of DNA.

Elongation

Once the DNA Polymerase has attached to the original, unzipped two strands of DNA (i.e. the template strands), it is able to start synthesising the new DNA to match the templates. It is essential to note that DNA polymerase is only able to extend the primer by adding free nucleotides to the 3' end.

One of the templates is read in a 3' to 5' direction, which means that the new strand will be formed in a 5' to 3' direction. This newly formed strand is referred to as the Leading Strand. Along this strand, DNA Primase only needs to synthesise an RNA primer once, at the beginning, to initiate DNA Polymerase. This is because DNA Polymerase is able to extend the new DNA strand by reading the template 3' to 5', synthesising in a 5' to 3' direction as noted previously.

However, the other template strand (the lagging strand) is antiparallel, and is therefore read in a 5' to 3' direction. Continuous DNA synthesis, as in the leading strand, would need to be in the 3' to 5' direction, which is impossible as we cannot add bases to the 5' end. Instead, as the helix unwinds, RNA primers are added to the newly exposed bases on the lagging strand and DNA synthesis occurs in fragments, but still in the 5' to 3' direction as before. These fragments are known as Okazaki fragments.

Table 4.2. A Process Description (continued)

Termination The process of expanding the new DNA strands continues until there is either no more DNA template left to replicate (i.e. at the end of the chromosome), or two replication forks meet and subsequently terminate. The meeting of two replication forks is not regulated and happens randomly along the course of the chromosome. Once DNA synthesis has finished, it is important that the newly synthesised strands are bound and stabilized. With regards to the lagging strand, two enzymes are needed to achieve this; RNAase H removes the RNA primer that is at the beginning of each Okazaki fragment, and DNA Ligase joins fragments together to create one complete strand.	**Using each stage name as a heading, the author provides details about the three stages.**

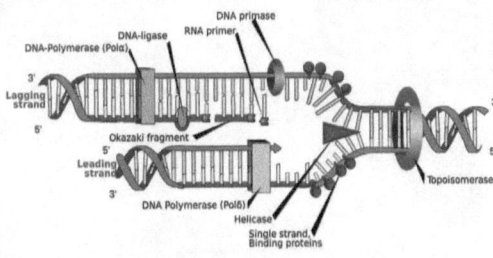

Fig 2.0 – Diagrammatic representation of DNA replication (By LadyofHats Mariana Ruiz [Public domain], via Wikimedia Commons)

Clinical Relevance—Sickle Cell Anaemia Sickle Cell Anaemia is an autosomal recessive condition which is caused by a single base substitution, in which only one base is changed for another. In some cases this can result in a 'silent mutation' in which the overall gene is not affected, however in diseases such as Sickle Cell Anaemia it results in the strand coding for a different protein.) In this case an adenine base is swapped for a thymine base in one of the genes coding for haemoglobin; this results in glutamic acid being replaced by valine. When this is being transcribed into a polypeptide chain the properties it possesses are radically changed as glutamic acid is hydrophilic, whereas valine is hydrophobic.	**Finally, the author discusses what clinical results occur when the DNA replication process goes wrong.**

Table 4.2. A Process Description (continued)

This hydrophobic region results in haemoglobin having an abnormal structure that can cause blockages of capillaries leading to ischaemia and potentially necrosis of tissues and organs – this is known as a vaso-occlusive crisis.

These crises are typically managed with a variety of pain medication, including opioids and NSAIDs depending on the severity. Red blood cell transfusions may be required in emergencies, for example if the blockage occurs in the lungs.

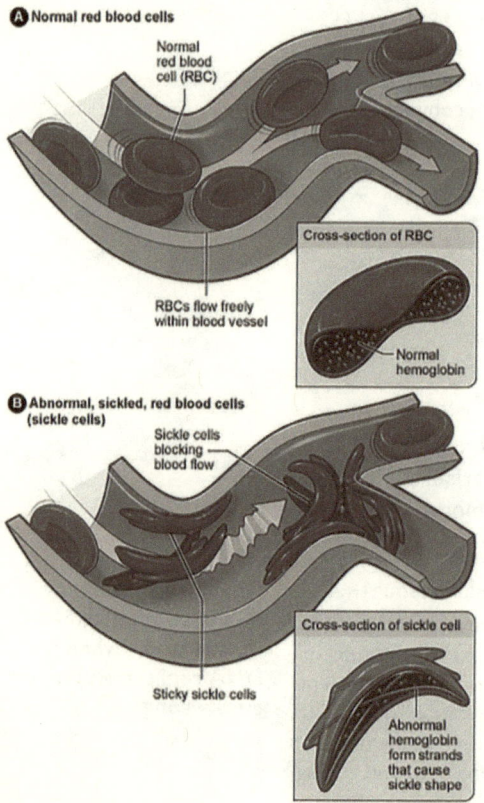

Fig 3.0 – The difference in structure between normal red blood cells, and those affected by sickle cell disease. (By The National Heart, Lung, and Blood Institute (NHLBI) [Public domain], via Wikimedia Commons)

Conclusion

In this chapter, I have introduced technical definitions and descriptions and discussed how to draft them. Technical definitions provide clear and accurate explanations of important terms and concepts. Technical descriptions inform readers of a mechanism or process and enable readers to take appropriate actions. They are not only common elements of technical documents, but also the backbone of effective technical communication.

Works Cited

Butterfield, Andrew, et al. *A Dictionary of Computer Science*. 7th ed., Oxford UP, 2016.

Center for Disease Control and Prevention. *About COVID-19*. https://www.cdc.gov/coronavirus/2019-ncov/cdcresponse/about-COVID-19.html. 16 June 2020.

Center for Disease Control and Prevention. *What is Autism Spectrum Disorder?* https://www.cdc.gov/ncbddd/autism/facts.html. 2 Aug. 2020.

Center for Disease Control and Prevention. *Facts about Developmental Disabilities*. https://www.cdc.gov/ncbddd/developmentaldisabilities/facts.html. 2 Aug. 2020.

Computer Hope. *Emoji*. https://www.computerhope.com/jargon/e/emoji.htm. 3 Sep. 2019.

Computer Hope. *Keyboard*. https://www.computerhope.com/jargon/k/keyboard.htm. 2 Aug. 2020.

Dovey, George. "DNA Replication." *TeachMe Physiology*, https://teachmephysiology.com/basics/cell-growth-death/dna-replication. 2 June 2020.

Encyclopedia Britannica, *Britannica Global Edition*. Britannica, 2016.

Lewis, Holden. "*Fixed-Rate Mortgages: What You Need to Know*." *NerdWallet*, https://www.nerdwallet.com/blog/mortgages/fixed-rate-mortgage-101. 2 June 2020.

Merriam-Webster. *The Merriam-Webster Dictionary*. Merriam-Webster, 2019.

Oxford UP. *Oxford English Dictionary*. Oxford UP, 2001.

Taylor, Mia. "*What Is a Mortgage*." *Bankrate.com*, https://www.bankrate.com/mortgages/what-is-mortgage. 2 Mar. 2020.

Understood Team. "*Text-To-Speech Technology: What It Is and How It Works*." *Understood*, https://www.understood.org/en/school-learning/assistive-technology/assistive-technologies-basics/text-to-speech-technology-what-it-is-and-how-it-works. 14 Sep. 2023.

United States Patent and Trademark Office. *A Guide to Filing a Design Patent Application*. https://commons.wikimedia.org/wiki/File:Fully_assembled_view_and_exploded_view.jpg. 2 Aug. 2020.

U.S. Energy Information Administration. *Renewable Energy Explained*. https://www.eia.gov/energyexplained/renewable-sources. 2 Aug. 2020.

Wallender, Lee. "*What Are Laminate Floors?*" *TheSpruce.com*, https://www.thespruce.com/what-is-laminate-flooring-1821619. 9 July 2020.

Teacher Resources

Overview and Teaching Strategies

Technical definitions and descriptions can be addressed in several ways in an introductory technical communication course. Instructors of such courses can incorporate technical definitions and descriptions in various kinds of projects, including but not limited to user manuals, product documentation, scientific publications, and Web content.

Instructors should follow the structure of this chapter by first providing learners an overview of technical definitions and descriptions, then analyzing the various types of definitions and descriptions, and ultimately practicing how to draft them. Because technical definitions and descriptions vary greatly by document genre, context, and audience goals, excellent examples that are contextualized can effectively illuminate this knowledge. Likewise, when students practice drafting technical definitions and description, such practices are best done in specific contexts of use. Because technical definitions and descriptions often come as part of a technical document — and are shaped by the broader goal of the document — instructors can embed relevant assignments in the creation of whole technical documents such as user manuals and product documentation. These whole documents provide rich contexts and defined audience that are ideal for practicing technical definitions and descriptions.

When asking learners to draft technical definitions and descriptions, instructors can consider the following activities:

- Begin with an analysis of audience and context. Instructors can help learners approach technical definitions and descriptions not as still dictionary entries, but as rhetorical messages that are aimed at audiences and contexts. Guide learners to investigate the intended audience and goals and apply concepts and strategies from this chapter.
- Gather subject-matter information. Instructors can help learners become skilled at gathering information from documents and consulting with subject-matter experts.
- Draft the technical definition or description. Instructors can guide learners to select the appropriate type of definition or description and apply appropriate strategies that respond to the given audience and context. It is also helpful to ask learners to try different strategies and compare how each strategy works with the context.
- Test with audiences. If conditions allow, it could be fruitful for learners to test their drafted technical definitions and descriptions

with target audiences. This process often allows learners to discover issues that they can't otherwise see and improve their ability to meet audience goals.

In summary, instructors can use this chapter to teach learners why technical definitions and descriptions matter, what they are, and how to create them. The key is to contextualize technical definitions and descriptions and link concepts from this chapter with specific contexts and audience goals.

Discussion Questions

To help learners understand technical definitions and descriptions, consider the following questions when analyzing or drafting these definitions and descriptions:

1. What technical definitions and descriptions do you encounter in everyday life? Where and how are they provided? In what contexts do you use them?
2. Who is the intended audience of a technical definition or description?
3. What are the primary goals of your audience in using a given technical document?
4. What do your audiences already know about the concepts to be defined and mechanisms or processes to be described?
5. What widely held views or misunderstandings, if any, exist in your audiences?
6. What resources are available on the subject matter?
7. What strategies are the most appropriate for defining a technical term or describing a technical mechanism or process?

These questions can be used to analyze examples of excellent and poorly written technical definitions and descriptions; they are particularly helpful when learners draft definitions and descriptions from scratch. The key is to help learners think about many factors that technical communicators must attend to when drafting technical definitions and descriptions and link these factors with the concepts and strategies from this chapter.

5 Let's Party: Composing a Review of the Literature on a Technical Topic

Daniel P. Richards

If you're reading this essay, then you most likely have a daunting task ahead of you: writing a literature review.[1] Rest assured that you are not alone in your apprehension or lack of excitement at the prospect laid in front of you—most writers feel the same. Parties have never been thrown for a literature review. No champagne bottles have been uncorked in its name.

It is likely that your experience in writing literature reviews to this point has been when composing that section of your research paper where all the secondary sources you found belong. You have located the requisite five sources for your teacher and now need to cite and summarize them. You dump them in the boring space of the document that exists after your clever introduction but before your savvy argument. That space might be more than boring; it might be treacherous, a place where all the MLA and APA errors live and where you feel right out of your league in being able to aptly summarize complex scholarship.

Yet, rest assured, a literature review remains a vital ingredient in the effectiveness of a given piece of writing, and it might even surprise you that literature reviews serve a critical function in technical and professional workspaces as well. Research does not end once a degree is conferred. A literature review can take many forms and can be found in a wide array of academic, scientific, technical, and workplace documents spanning all fields and disciplines. Sometimes it is the full document; sometimes it is but part of a document. From essays in philosophy to journal articles in oceanography to grant applications in microbiology to proposals in public policy to informal reports in social media marketing to business pitches in accounting, each have a type of literature review component that are connected more by *function* (what it does) than by *form* (how it is organized).

---

[1] This work is licensed under the Creative Commons Attribution-NonCommercial-NoDerivatives 4.0 International License (CC BY-NC-ND 4.0) and is subject to the Writing Spaces Terms of Use. To view a copy of this license, visit http://creativecommons.org/licenses/by-nc-nd/4.0/, email info@creativecommons.org, or send a letter to Creative Commons, PO Box 1866, Mountain View, CA 94042, USA. To view the Writing Spaces Terms of Use, visit http://writingspaces.org/terms-of-use.

That is, the *form* can take many shapes: A literature review can be a three-page section of an academic essay, yes, but it also can be a one paragraph overview of common platforms used by other companies in their social media marketing. It can be a background research section of a grant application for a municipal project. The *functions*, however, by and large stay the same. The goal of this essay is to examine what these functions are and, in doing so, present the case that a literature review is not supplemental but foundational to any piece of writing—including those in more technical documents.

So, actually, you know what? Let's throw literature reviews a party, after all. And let's do so by way of an acrostic (a type of verse in which the first letters of each line form a word) of five functions of a literature review, which we'll address in order, and which end with practical guidance. In terms of its functions, a literature review:

- Participates in a Conversation
- Adjudicates Sources
- Refines for the Target Audience
- Transmutes into Trust
- Yokes You to Others

Literature reviews deserve as much, given their bad rap, indeed their pivotal but thankless role in writing success. And it might make the prospect a little clearer and, dare I say, exciting.

Participates in a Conversation

Alright, first things first: "literature" doesn't mean what you might think it means. In the context of a "literature review" the term means much more than just literary texts, especially those covered in English and, well, *literature* classes. In academic and workplace contexts, "reviewing the existing literature" on a topic means to account for the work on your given topic that has come before you in this current moment in time. This work can be books, of course, but also journal articles, websites, projects, funded grants, white papers, or any document that discusses the best practices of a given process. To review the literature is to acknowledge the predecessors that came before you. It is to participate in the larger conversation around the topic you are researching.

This idea of participation is key. While it may seem like secondary research is static and somewhat disconnected from your current writing project in front of you at this very moment, discovering what research, practices, and perspectives have been established before you actually connect your work to a larger system of ideas. Discovering what has already been done by others does not subtract from the value of your work. It adds to it by linking it directly to an ongoing conversation in which you are seeking to be a participant.

Imagine showing up hours late to a party and your friends are having an unexpectedly vigorous debate. As you approach closer to the loud noise of your fiery friends you notice that they seem to be debating whether or not a beloved local coffee shop's new app allowing for online ordering is a good idea—a contentious topic if there ever was one! As you reach the circle, you apologize for interrupting with your lateness and vocalize, as a regular patron of the local coffee shop and former coffee barista yourself, that you think the new app is ridiculous and will take away the essence of the local coffee shop experience.

You even surprise yourself at how riled up you are about this topic and continue on with examples from your time as a coffee barista where you would get to know and even anticipate orders of regular customers from the moment they walked in the door. You gave examples of Julia's double mocha at eight in the morning, or Ahmad's late afternoon triple espresso on his way to his late shift. You fervently insist that getting to know customers personally is part of the coffee shop experience, and that the time spent chatting while the customer is waiting in person is a vital social event to our culture! We need to retain these coffee shops to retain our sense of community, you exclaim! The faces of your friends slowly take puzzled forms; a few even glance at each other, polite but confused. "What?" you ask. "Am I being wild? Do you all disagree with me? I'm not being radical here!"

"No, you're not," your closest friend, and now spokesperson for the group, calmly responds. "It's just that we weren't talking about local coffee shops generally and their social functions. We were talking about how everything in this world is just speeding up, and how no one has any patience for anything anymore. It's go-go-go all the time, and this new app is just one example. This conversation actually started with social media. We were a bit thrown off by your, um, statements because your story about your time as a barista was pretty deep in the weeds about coffee shops and not at all about the death of patience in our society. So, it's kinda related, but, like, not really, right?"

How embarrassing. You completely misread the conversation. You want to go back home, or at least sink into a hole in the floor, but you simply apologize and tell yourself you're going to learn how to listen better and be less opinionated all the time. You're quiet for the rest of the night, which is probably for the best.

The party rule you broke was the unstated condition of participation in conversation: that you listen and acknowledge and *then* speak. You might feel like you have a grasp of a conversation by the last one or two things said, but without the full understanding of the conversation's starting point, not to mention the opinions of others that have been stated before your arrival, your vocalization is not really participation in the truest, most honest and productive sense of the word. You're making noise and people hear it, but it is not connected and, therefore, not really useful. Or persuasive. Or meaningful. By analogy, the literature review helps satisfy this condition. It shows that you have been paying attention, you understand the origins of the conversation, and you acknowledge and value the words of others above or at least before your own.

To improve this function of your literature review, you'll need then to conduct a thorough review of all the work that has been done and is currently being done on your topic. This ensures a more seamless entry into participation and also ensures that you are not repeating the work of others. It shows you are "listening" before speaking. This process of review must be diligent, even painstakingly so, depending on the level of depth you are seeking to achieve.

The reader has absolutely no obligation to do any research for you. You will, of course, have to draw the line somewhere—you can't research everything *ever*. So, seeking the guidance of your professor or a subject-specific librarian would be wise to help scope the breadth of the project. The type of reviewing and researching you'll be doing and the places you'll look will depend on the piece of writing you're working on. And each piece will have different places to start and different times to stop.

Practice

Find everything you can on the topic, using resources around you like a librarian or professor to point you in the right direction. Create a spreadsheet—a running inventory of sorts—to archive your finds (Table 5.1).

Creating a table ensures that (a) you don't have 21 browser tabs open with important information and then lose them when your internet crashes and (b) you are thinking about how each piece of information you find—could be books, journal articles, web pieces, blog posts, or software applications—connects to your research. They could support it, challenge it, facilitate it—whatever. But listening to as many folks as you can and learning (and remembering!) their positions on a topic is one of the best ways to have a great time at a party and avoid any social *faux pas*. Off you go to dig into the research.

Table 5.1. Source Spreadsheet for Literature Reviews

Author	Title	Publication	Link/ DOI	Reliability Rating (1 to 5)	Main Argument/Findings	Connection to Your Project
...	...	...	...	...	...	...

ADJUDICATES SOURCES

Whew! That was a lot of sources. You have currently . . . let's see . . . 5 . . . 10 . . . 17 sources! Great work. But your professor only asked for a few. Now what? Need you include all of them in your literature review? Short answer: No. *But I want the professor to know how much research I've done*, you retort. Understandable, but another key function of the literature review is adjudication—or, judging *which* sources are most relevant to include and *why*, essentially sifting through your list and transitioning from indiscriminate collection of sources to a more focused curation of the most credible and relevant sources for the constraints, genre, and purpose of your piece of writing. This takes time, experience, and deliberation—much like it would for any astute judge.

The process of adjudication serves the function of scoping your topic. So, when I say that you curate your sources according to constraints, genre, and purpose this means that you narrow down what you'll be including by thinking about how much space you have, who your audience is, what the expectations are for type of document, and what it is you are trying to accomplish. Sometimes literature reviews are one paragraph in a report, and sometimes they are 10 pages in a research paper. Sometimes the source is a journal article and sometimes it is a software application. Again, the form differs widely but the function of focusing and framing your topic remains across the board.

As an example: Imagine you just got hired for an entry level communications position at a local nonprofit environmental organization whose mission is to engage the public on issues around global warming and sea level rise. This is your dream job! You're on the communications team for a science-centered organization and can now put your English degree in technical writing to use. You're excited to funnel in your passion about protecting the earth and its residents to a real job. You're also interested to see how your academic degree will transfer into a real-world job context.

The development team contacts the communications team with an idea. They would like to develop their own sea level rise viewer that is interactive and that would allow folks along the coastlines to see what a certain amount of flooding or sea level rise would do to their home and when. They are sick and tired of trying to connect to residents by sharing scientific data; they want to help personalize the risk in the hopes of engaging residents who are at risk. They have some preliminary design ideas, and they have a good sense of what type of coding needs to go into such a GIS-focused tool. But they want to see what the current research is on these types of tools. They of course reach out to the communications team because this team consists of astute researchers with freshly minted degrees in the humanities.

You start accumulating sources in your nifty table (Table 5.2). During your research you start to realize that on this topic there are academic sources as well as actual tools (like applications and maps). Are these "sources," you wonder? Is this part of the "literature" that needs to be reviewed? In technical contexts, the answer is *yes*. When exploring technical topics, all sources are game, even if they don't fit as neatly into the nifty table.

You consider what sources would be useful for the development team to know about. You really engaged with the Rawlins and Wilson journal article, most likely because you have a knack for theory, but would this be helpful to review for the purpose ahead of you? The constraints of the informal report genre? The audience at hand? The Stephens et al. piece you also enjoyed and was much more practical and research based. But they focused on the user experience of sea level rise viewers, which seems farther down the project road for the developers, who have barely started prototyping their vision. You realize now that a source can be quite reliable, engaging, and connected but not of immediate relevance for your project. And that's OK—that's part of savvy rhetorical adjudication that focuses on the needs of the reader more than a window into the full extent of your research.

You then decide that for this "review of literature" you'll just be listing, summarizing, and synthesizing the actual tools themselves. You'll make a note in there about how there is ample scholarship on these tools, but for the purpose of this project, and its scope, at this time you are choosing to focus on the technical sources.

Table 5.2. Populated Source Spreadsheet for Literature Reviews, Including Both Academic and Technical Sources

Author	Title	Publication	Link/DOI
Sonia H. Stephens, Denise E. DeLorme and Scott C. Hagen	Evaluating the Utility and Communicative Effectiveness of an Interactive Sea-Level Rise Viewer Through Stakeholder Engagement	*Journal of Business and Technical Communication*	https://doi.org/10.1177/1050619155 73963
Reliability Rating (1 to 5)	**Main Argument/Findings**	**Connection to Your Project**	
5	The authors discuss the implications of this study for visual risk communication and make recommendations for others developing similar interactive data visualization tools with audience input.	This is research on the user experience part of the tool, and not its original development.	
Author	Title	Publication	Link/DOI
Jacob D. Rawlins and Greg D. Wilson	Agency and Interactive Data Displays: Internet Graphics as Co-Created Rhetorical Spaces	*Technical Communication Quarterly*	https://doi.org/10.1080/10572252.2014.942468
Reliability Rating (1 to 5)	**Main Argument/Findings**	**Connection to Your Project**	
5	Bit complex and theoretical, but the argument seems to be the concept of agency is a useful perspective to interpret the way users interact with a variety of online graphical tools.	This could help with placing this project into an academic one that is more philosophical.	
Author	Title	Publication	Link/DOI
NOAA	Sea Level Rise Viewer	https://NOAA.gov	https://coast.noaa.gov/digital-coast/tools/slr.html
Reliability Rating (1 to 5)	**Main Argument/Findings**	**Connection to Your Project**	
5	Hmm. That sea level rise risks should be visualized? More by implication, really.	This resembles a type of tool that I think the development team wants to create.	

Table 5.3. Populated Source Spreadsheet for Literature Reviews, Including Both Academic and Technical Sources

Author	Title	Publication	Link/DOI
Climate Central	Risk Zone Map	Surging Seas?	https://ss2.climatecentral.org

Reliability Rating (1 to 5)	Main Argument/Findings	Connection to Your Project	
4?	Like NOAA, the argument is the same.	This resembles a type of tool that I think the development team wants to create—but more attainable since they are also a nonprofit, like we are.	
...	...	...	...

Practice

You'll need practice in "summarizing" technical sources. Consider adapting the table from the previous sections to highlight information that is more relevant to technical sources (e.g., swapped out DOI and reliability for functionality and user base). I've input the NOAA source as an example (Table 5.3).

As most technical writers find out, what constitutes useful sources will constantly change according to purpose and will extend far beyond musty books and full-text periodicals. (We're two functions in and you haven't even started writing yet!)

Table 5.4. Source Spreadsheet Adapted for Technical Sources

Agency	Type	Link
NOAA	Government	https://coast.noaa.gov/digitalcoast/tools/slr.html
Origination Date	**Functionality**	**User Base**
Not sure. Last modified August 2020.	Identify address and use sea level gauge (inter-active).	Primarily coastal planners but also the public.
...	...	...

Refines for the Target Audience

One of the finer aspects when you get to the actual composing stage of things is making decisions about how much detail to share and what level or language to use. In the previous case, now that you've decided about the *types* of sources you'll be including, you now need to ascertain how you'll be framing the review and how *deep* to go into them. Should you spend one sentence summarizing a source or 600 words? That depends on the genre and the amount of space you have. The most astute literature reviews also attend—not surprisingly—to the audience's needs; the language and assumptions about what they know need to be as tailored as possible. The research now needs to be refined, like that fancy wine filter thingy that elegant folks use at wine and cheese parties. I'm sure it has a name.

So, for this you'll really have to think hard about what you know about your audience so you can further improve their reading experience. If the development team are the "techy" people in the organization, what about them? Are they computer scientists? Data scientists? Web designers? Coders? Do they have college degrees? What are the types of technology they typically deal with? You need to find out as much as you can about them.

I see this issue all the time in academic papers. I'll assign a journal article for students to read and when it comes time to cite or reference that work in their paper, they assume the audience knows exactly who the authors are and what the field is. Even though I am the actual reader, assignments that ask students to "write to a more public audience" still suffer from this assumption. It'll often go like:

> As Stephens et al. (2015) write, the field of UX is an important consideration of SLR viewers . . .

Not wrong, necessarily—but misguided. First of all, would the public know who Stephens et al. are? Are they scholars? And of what? Have they written about this before? Are they emerging or renowned scholars? And, also, what are UX and SLR? Are those acronyms? When summarizing a piece for a literature review, you need to absolutely calibrate your level of writing and modify your assumptions about pre-existing knowledge of the readers.

There is a big difference, then, in writing a summary of the NOAA tool for a group of data scientists than for web designers. For data scientists, you might summarize the NOAA tool as:

> NOAA's Sea Level Rise Viewer is part of their Digital Coast project, an online repository of oceanographic and marine data accumulated using LiDAR technology, tide gauges, climate projections, and historical trends data sets.

"Projections, "data sets," and "repository" would be recognizable language for data scientists, most likely, regardless of their area of expertise.

And it also highlights what they would be interested in, which are the data underlying the tool. A web designer audience, on the other hand, would most likely want to know what these tools look like and what web elements exist. It might go something more like this:

> NOAA's Sea Level Rise Viewer is part of their Digital Coast project, which seeks to engage a more public audience as well as city and coastal planners by visualizing sea level rise data in an interactive map that allows for personalization and exploration.

In this case, the web designers would want to know about target audiences and users as well as the visual aspects of the tool—essentially, how it would be used. Notions of audience, "personalization," and "exploration" would be familiar concepts to web designers.

Practice

But how do you get to know your audience so well? Sometimes it is useful to create an audience profile sheet. This sheet can be filled in based on what you and your team already know about the audience or by asking them directly. An audience profile sheet might ask the following questions:

- What are the job titles of the readers?
- What are their backgrounds (majors in college, etc.)?
- What are their areas of expertise (GIS, web development, research, etc.)?
- What do they currently know about this topic?
- What do they want to or need to know about this topic?
- What are they expecting this piece of writing to look like?

Exhausted yet? Well, yes, writing effectively after conducting copious amounts of research isn't for the faint of heart. But the next function should bring this party back to life! (Even if it does begin with an archaic word . . .)

Transmutes Into Trust

We should probably define transmute, eh? OK, Merriam-Webster (2020) defines it as such: "to change or alter in form, appearance, or nature and especially to a higher form." (Which is probably what we wanted to do after embarrassing ourselves at the party by sharing our opinion on the new local coffee shop app, am I right?) This notion of alteration or change in our context pertains to the fact that a useful, well-research, tailored literature review meets your readers' needs, yes, but it also imbues yourself as an author with a degree of trustworthiness. Conducting copious amounts of research, adjudicating it, and then adapting your summary of said sources to your audience, whom you actively sought to learn more about, brings about the positive net result of the reader knowing what they need to know but now also trusting your judgment.

This is why literature reviews are more than just the banal task that feels like a formal requirement in papers. Done well, literature reviews are just as persuasive as the argument or position or proposal you lay out because, rhetorically speaking, *ethos* (your credibility and character) matters. Perhaps you've had this experience where two people have presented the same argument or provided the same advice to you, in the exact same form and construction, but because it came out of one person's mouth and not the other, it spurred you to change or consider it—most likely because you trust one person over another. The source of an idea matters, and we have both conscious and unconscious ways we trust people and their ideas. If a well-written literature review can increase the degree of trust in your reader, then why not do this work? You might have a savvy argument but if you cannot show the reader that you are aware of the research on that topic that came before you, then why would someone believe you?

Let's say, for example, that you have an internship with your local city government in their office of elections. You notice that on their website they include flyers educating college students on how to register to vote. These flyers are poorly designed; you ask your director if you can take a shot at revising them knowing what you know about usability, accessibility, rhetoric, and document design. Your director agrees, but she'll need first to see an informal proposal of why this work needs to be done before she can approve such a time-consuming project and take you away from other tasks. In essence, she needs to know that she can trust that what you're doing is worthwhile.

You begin work on the proposal, which has the fairly clear purpose of persuading your director that design is important in voter documentation materials. But *how* to communicate that purpose? You get the sense that your director is content with having the information on flyers, regardless of form or design. *The information is accurate after all, so why should we spend any time on revising them, on redesigning them*? All fair questions, to be sure.

You begin to realize that the type of research your director will need to see is research on the positive impact on well-designed voter documentation materials—ranging from brochures, to flyers, to ballots. She knows accurate voter documentation materials are important, of course, but perhaps not that much about the importance of well-designed ones. More than that, she'll need to trust that you know what you're talking about. So, this is less about overwhelming your director with evidence after evidence but more about eliciting trust by establishing shared interest. And this can happen even within the context of a low-stakes informal proposal. You begin to cull the resources from your undergrad courses in writing and design, knowing that a well-researched proposal will give your director the faith and trust required to have you move forward with this project.

Practice

One strategy is to reach out to someone that you yourself trust on the given topic and have them review what you've written. A brief peer review from an expert in that field will help ensure you are depicting the field or topic well.

Yokes You to Others

Another odd, archaic word? Really? Yes, really. Back to the dictionary: Yoke is defined in many senses. The first is: "a wooden bar or frame by which two draft animals (such as oxen) are joined at the heads or necks for working together" ("Yoke"). Are there oxen at this party? Maybe. Maybe not. But the essence of it is there, and more evident in the last primary definition: "a clamp or similar piece that embraces two parts to hold or unite them in position." That should be clearer. The literature "yokes" you to others by uniting your voice with theirs, even if it is in agreement or disagreement.

This is different than the first function of participation. Yoking yourself to others means the literature review lets readers know the individuals and texts and projects and ideas you think are important and should act as a lineage for your current piece of writing. In some cases, you don't have much of a choice in a key scholar or practitioner to cite. It is hard, for example, to talk about design and user experience without at least tipping your hat to Donald A. Norman, author of *The Design of Everyday Things*. And in most cases being yoked to Norman's work is in your best interest, since it connects you to a major text and instills the audience with your credibility for knowing who the key people are in the field of study at hand.

In other cases, you have more leeway in choosing the folks you'd like to unite with. There is power in the literature review because by selecting one source or another, you are telling the reader: "This is how the story goes and who has written it so far." There is power in these decisions. In this sense you are not only participating in a conversation and selecting the sources, you are, theoretically, in some cases, actually becoming a part of the collective of thinkers who care about this topic. You are no longer a passive spectator but rather an active participant, yoked to folks who also care. You are oxen tilling fields of knowledge.

Practice

Select two sources that are closely related, either in argument or method. Review the sources they reference or cite in their own work, and ask: Are they the same? Different? Why or why not?

Conclusion

Before we finish, you should know that there is a popular concept created by famous rhetoric scholar Kenneth Burke called "the unending conversation." He uses the metaphor of a parlor to help describe how argument works, and I think it is a useful way for you as a student to understand literature reviews as well:

> Imagine that you enter a parlor. You come late. When you arrive, others have long preceded you, and they are engaged in a heated discussion, a discussion too heated for them to pause and tell you exactly what it is about. In fact, the discussion had already begun long before any of them got there, so that no one present is qualified to retrace for you all the steps that had gone before. You listen for a while, until you decide that you have caught the tenor of the argument; then you put in your oar. Someone answers; you answer him; another comes to your defense; another aligns himself against you, to either the embarrassment or gratification of your opponent, depending upon the quality of your ally's assistance. However, the discussion is interminable. The hour grows late, you must depart. And you do depart, with the discussion still vigorously in progress (Burke 110-111).

Certainly, my use of "P-A-R-T-Y" relies on Burke's inspiration. The similarity of this description to mine in the first function, *participation*, wherein you argue about the social function of local coffee shops is obvious. The main difference I wanted to make is that in Burke's construction, there is success. In ours, there is not.

Imagine, though, if I did not cite Burke here? I'm sure even the professor who assigned you this chapter was thinking at some point, "Hmmm . . . odd how Dan doesn't mention Burke's parlor at all here even though he is borrowing heavily from it . . . Especially since his Ph.D. is literally in rhetoric." And that's just it. While not in the form of a literature review, my reference and acknowledgment of Burke saved face a little. It bolsters my *ethos* in the fields of rhetoric and technical writing, and also gives you, the reader, trust that I am at least giving guidance of some value, since I have yoked myself to one of the forbearers of modern rhetoric.

It might feel as though writing is a solitary activity because so much of it is done alone. But writing is always a social activity—writers are never alone (Ede). It may seem as though your argument about the value of a government policy, or your analysis of a text is your own. But the literature review portion of a document is the main connector of your work to others'. Without acknowledging the arguments that came before you, or the research conducted prior to your own, or how you have been influenced by—even indebted to—other ideas from other writers, your argument goes nowhere. Use this as a mindset while you are finding sources. Researching for sources is not a banal task but a basic guideline for participating in something bigger.

And that, to me, is worth popping a bottle over. Cheers to literature reviews!

Works Cited

Burke, Kenneth. *The Philosophy of Literary Form*. Louisiana State University Press, 1941.

Ede, Lisa. "Writing as a Social Process: A Theoretical Foundation for Writing Centers?" *The Writing Center Journal*, vol. 9, no. 2, 1989, pp. 3-13.

Norman, Donald A. *The Psychology of Everyday Things*. Basic Books, 1988.

"Yoke." *Merriam-Webster Dictionary*. https://www.merriam-webster.com/dictionary/yoke.

Teacher Resources

Overview and Teaching Strategies

Instruction on composing literature reviews typically happens near the beginning of a writing project. As such, this chapter can be integrated into the early phases of student writing either as a theoretical framework to introduce students into the idea of a literature review or as a heuristic to get students started on gathering, curating, and annotating sources that are to be potentially relevant to their piece of writing—or both! There is a procedural element to this chapter that might allow for instructors to use it as a guide over the course of one or two weeks at the outset of an assignment, as students transition from finding sources to evaluating them to curating them.

Given my argument that literature reviews are an integral part of many genres, from academic papers to informal proposals, this chapter can be applied to most any genre covered in a technical writing classroom. That said, individual instructors will need to supplement this chapter with specific guidance on the particularities of the nature of literature reviews for each genre. Issues such as length, style, tone, and depth of "literature reviews," broadly defined, are determined by genre, industry, and organization. Integrating these specific features into the larger instruction on literature reviews will be most helpful.

The tables used in this chapter also might lend themselves well to collaborative work, where groups of students are tasked with finding and synthesizing their researched sources into a centralized document.

Discussion Questions

Consider supplementing this chapter with the following questions:

1. How have you been taught to write literature reviews in the past? Was it always in the context of academic papers?
2. Where else do you see the function of literature reviews in contexts not discussed here? Do you see overviews of existing content take shape in, say, YouTube videos? If so, how?
3. In a small group, discuss how literature reviews vary by major, field, or discipline. What differences do you see? And why do you think those differences exist?
4. Composing an effective literature review requires handling a good deal of information. What are some ways you can manage your content and information in an efficient way?

6 Stronger Together: Collaborative Work in the Technical Writing Classroom

Laurence José

Group work is hard.[1] It requires careful planning, negotiation, and trust. Most of us, as students and teachers, have been in a situation when we were assigned a collaborative project and thought how easier it would be to do it alone. Collaborative writing projects are no different. Before even considering questions related to work distribution and fairness, it is very tempting to think of writing as a solitary act that a group setting slows down and makes more difficult. What we often forget though is that collaboration can be an asset by giving us more ideas to draw from to generate and shape the content for the audience. It also gives us opportunities to participate in bigger projects that could not be completed by one person alone and it helps to emulate workplace writing experiences. Collaborative writing is indeed a key component of many professions and a core competency in technical writing. Learning how to do it well does not just make for higher quality class projects, it can also offer space for students to work with clients and, consequently, become a line on a résumé to transition into post-academic contexts.

College classes often involve some level of collaborative work. Technical writing courses are no different. In fact, collaboration has long been identified as one of the key components of technical and professional writing courses (Allen and Benninghoff). Yet, many of us have experienced times of anxiety and even dread when we mark the days on the class calendar that precede the start of the group project. From communication issues to fear of unequal workload distribution or scheduling difficulties, let's be honest, there are many reasons for being leery about group work.

Because we tend to think of writing as a solitary act, collaborating on writing projects may seem even more complicated. How can one create

---

1 This work is licensed under the Creative Commons Attribution-NonCommercial-NoDerivatives 4.0 International License (CC BY-NC-ND 4.0) and is subject to the Writing Spaces Terms of Use. To view a copy of this license, visit http://creativecommons.org/licenses/by-nc-nd/4.0/, email info@creativecommons.org, or send a letter to Creative Commons, PO Box 1866, Mountain View, CA 94042, USA. To view the Writing Spaces Terms of Use, visit http://writingspaces.org/terms-of-use.

and produce content as a team? How should the project be handled so that the final product is cohesive and reflects everyone's input? How can one make sure one group member does not end up handling most of the workload? And, perhaps most importantly, how can one use collaboration to their advantage to make the end product better than a solo-executed project? These are all questions that this chapter addresses. Specifically, this chapter provides strategies not only for developing content as a group but also for making a collaborative writing project meaningful and, yes, even fun. Really, what is the point of collaboration if a project is easier to complete alone?

What is there to Gain?

The first step toward a successful collaborative writing experience is understanding WHY it matters and WHAT we can gain from it. The idea behind this is simple: once we understand the relevance of a task, we are more likely to be motivated and engage with it in a productive way.

The main reasons for learning how to write collaboratively are twofold: first, it is a way to make the writing more authentic and reflective of what it does and is outside of the classroom; second, it is a way to develop tangible skills for your résumé.

Reason 1: Connecting Writing to its Realities Outside of the Classroom

Most writing performed in the workplace involves more than one person. As noted by scholar Kristin Woolever, "collaboration is the way work gets done today" (141).

Whether it is developing a report, a grant proposal, content for a website, or a series of instructions, professional writing generally involves input from different people at different stages of the process. Examples of such people can be:

- A subject matter expert (SME) whose expertise lies with a specific technical product or process and provides you with information for a technical report on a particular product or for instructions for a specific piece of software.
- A team member with whom you collaborate on the research and development of a grant proposal.

- Your boss for whom you are drafting an official email, letter, or any other document that will be sent within the company or go out to clients.

The prevalence of collaborative writing in the workplace also explains why it is used frequently in the technical writing classroom.

For example, speaking of the technical writing/communication classroom, scholar Liz Lane describes it as "wildly collaborative, in that we prepare our students to collaborate and test project management skills for eventual implementation in their professional lives beyond the classroom" (32). In their 2021 book chapter on "The rhetoric, science, and technology of 21st century collaboration," scholars Ann Hill Duin, Jason Tham, and Isabel Pedersen describe collaboration as an imperative for today's technical communicators and emphasize its importance in the classroom as means to prepare students for the workplace. In other words, think of collaboration not only as one of the key topics of technical writing but also as an activity that emulates the kind of writing you will encounter in your future professional lives.

Understanding Collaboration Types

Depending on the situation and the project, the collaboration will occur at different moments of the writing process and have different functions. It may also be more or less involved, ranging from punctual collaborative moments to projects that are entirely collaborative from the beginning to the end.

For example, going back to our first example of working with a subject matter expert (SME), let's say you are tasked with developing instructions for a specific software. The SME may work with you during the planning phase of your document and again during the testing and revision phase. In this case, the SME acts as a source for the content as well as an editor, perhaps in an unofficial capacity but editor nonetheless since their knowledge is what will guide the development of your document.

If you write a grant in a team environment (cf. previous example 2), you probably will be working as a team throughout the whole process. Depending on the team members' area of expertise, one member may analyze the requirements for the grant (grant analyst), another may research the grant's target audience and their values (researcher and fact checker), while another one may take the lead on developing the content of the proposal (the grant writer and editor).

Though it may be less evident to think of it as such, the writing of a correspondence document sent by a company to its clients also involves a form of collaboration (cf. earlier example 3). Even in this case, you are not writing alone and for yourself but with and for someone else. Your collaborator's role may then be limited to being a gatekeeper (that is, the person approving and signing off the document to be sent/published) but the fact remains that they are part of the writing process. The document that you produce speaks for and reflects someone else's purpose besides your own.

Making it Matter

The three previous examples are good reminders of the inherently collaborative aspect of writing in the workplace. In this regard, collaborating on writing projects prepares you for what writing is and does outside of the classroom. It makes your coursework more tangible and authentic by connecting it to your future professional life. Keeping in mind this bigger picture can also help you set your own learning goals. For instance, remembering how a project helps you to gain practice with the kind of communication you will encounter in your future career will make it easier to focus on what you want to take away from a project. This may also give you an extra source of motivation, besides earning a good grade.

Reason 2: Developing Tangible Skills for Your Résumé

Beyond these immediate writing-specific learning outcomes, working on a collaborative project is also a great practice for developing the kinds of soft skills that employers value.

Developing Soft Skills

Collaboration can be a catalyst for the development of soft skills such as being a good listener, showing empathy, learning to see situations from different perspectives, being able to negotiate different points of view, and being able to problem solve and resolve conflicts. For instance, a solo writing project can make it difficult to think about how a specific audience may interpret your text or react to your ideas. But when you develop content with a team, you must confront your ideas with others early in the process, and thus gain experience with editing as well as listening and negotiating. If such skills are sought after by employers, it is not always easy to provide evidence on your résumé for how and why you possess them. Listing a group project as evidence and example can do just that.

Working with Outside Audiences

Collaboration may also give you opportunities to work with communities and clients outside of the classroom. For example, if your technical writing course entails a service-learning component (that is, a project with an off-campus community partner), your team may be paired with a client such as a small nonprofit and be tasked with developing content for their website. Research has shown that class projects involving partners outside of the classroom are stronger motivation factors for collaboration than projects that remain inside of the classroom (Rebecca Pope-Ruark et al.). This relationship is yet another way of thinking about the WHY of collaboration in the technical writing classroom: besides giving you easy ways to get feedback on your work and to strengthen your communication and listening skills, it can also open up exciting possibilities for client projects that can be listed as experiences on your résumé.

So, when you are assigned this pesky group project, make it count and connect it to your long-term goals. This is not much different from any other class assignment, really. One of the challenges of class assignments is that they may feel artificial or disconnected from our immediate lives or future careers. How often did you write a paper, thinking "no one will ever care about this work except my instructor?" or "how does this connect to what I will do later?" Of course, this can affect your motivation. This is where thinking of an assignment beyond the immediacy of the classroom helps. It is a way to remember the WHY of the assignment and to emphasize WHAT you want to gain from it in the long run, even if in the end your instructor is your primary audience. Most importantly, it will make the work more meaningful and possibly ease worries you may have about it. For a collaborative project, always keep in mind how it will give you important skills for your résumé and future career. If you are not convinced, do a quick online search on "workplace collaboration" before starting your group assignment and see how many results you get and check what job recruiters say about it in your own field.

What is an Effective Team?

Once the relevance of collaborative writing is clear to you, the next step is to understand HOW to do it successfully. In other words, the question becomes: what makes a team effective?

Project Completion and Technical Skills Matter but They are Not Enough

It may be easy to equate team effectiveness with project completion. While submitting a project that is complete and on time is part of success, it is not all there is to it. Indeed, a project with one team member carrying out most of the work can hardly be considered a success or a model for future collaborations. Research has shown that communication and team cohesion are fundamental elements in a team's success (Lam; Murphy et al.). Even more so, lack of group cohesion and dealing with non-contributing members are the elements students mention most often as reasons for disliking group projects (Hall & Buzwell; Lam; Williams, et al.).

Depending on the class and the specifics of the project, you may be able to select your team yourself or you may be put in a team by your instructor. If the goal is to have everyone contribute, it may be tempting to think then that the most important element for a successful collaboration is a team whose members have the right technical skills for the project. Though important, it is (sadly) not enough to guarantee a successful collaboration. If it were, it would make collaboration much easier to plan since it would pretty much just be a matter of adding up the right technical skills. From experience, most of us know that it is far more complicated.

Imagine a project for which a team creates a series of instructions for a specific online application. A skill-based approach to the team's composition could involve questions such as:

- Who has experience with this specific app?
- Who has previous instruction writing experience?
- Who has information and document design experience?
- Who is a strong writer and editor?

While these questions must be discussed by the team at some point, they cannot be the only consideration. They should not even be the first questions asked during the planning phase. Indeed, even if a team is composed of members with all the appropriate technical skills, the team will not be effective if the members do not manage to work together as a cohesive group. Even if the end-product is completed and delivered, a project environment riddled with conflicts and tensions cannot be described as successful.

Ultimately, what makes a team effective cannot be measured by the addition of specific technical skills or even by the delivery of an end product.

Rather, it should be assessed by the ability of its members to collaborate and complete the project as a team. Differently put, what makes a team effective is the capacity to manage a project, to develop common goals, to get along and trust each other.

If equitable work distribution and effective contributions from all team members are key factors in a team's success, they are also the most difficult challenges to overcome. This also explains why the issue of non-contributing members, often referred to as social loafing, has become an area of research. For instance, scholar Chris Lam conducted a study that identified "communication quality and task cohesion" as key factors for reducing social loafing (454). Lam's conclusions and his focus on communication quality and team cohesion are on par with what Google discovered in its study on collaboration.

Google's Project Aristotle

Because collaboration is a crucial aspect of today's workplace, questions related to effective teamwork are of interest to many companies. The tech company Google, who is well known for its reliance and promotion of teamwork, launched a project in 2012 named Project Aristotle with the goal of determining what makes a team effective. The name Aristotle is a reference to the famous Greek philosopher and specifically to one of his famous quotes, "the whole is greater than the sum of its parts." The quote can be read as an evocation of what makes collaboration complex and the fact that a team's success is not simply determined by the individual skills of its members. Differently put, understanding collaboration requires understanding what makes the togetherness of a team successful. And this was precisely Google's goal for launching Project Aristotle.

Following two years of studying over 180 teams and intense data collection, the project identified 5 key characteristics and drivers for team effectiveness. In order of importance, the characteristics are as follow (Duhigg; Re:Work):

- Psychological safety: this refers to how safe a teammate feels to take risks without being considered ignorant or incompetent (Edmonson).
- Dependability: team effectiveness is also driven by members being able to rely and count on quality work from their teammates.
- Structure and clarity: an effective team establishes a clear structure for the collaboration and sets clear goals for the project.

- Meaning: in an effective team, each member finds a sense of purpose.
- Impact: team effectiveness is also driven by members identifying the "impact" that the work makes (Grant). Another way of thinking about this is to say: team effectiveness is driven by members' ability to identify the "so what" of the work.

One of the main and, perhaps surprising, findings of this study is that a team's effectiveness depends greatly on how safe team members feel. That is, for group work to be effective, team members must know that the team is a safe space and that they will not be mocked or rejected for speaking up and sharing their ideas. In many ways, Project Aristotle gives us additional and measurable data that confirms not only our experiences with group projects but also writing studies research results and the importance of communication quality (Lam; Wolfe).

WHAT ARE THE STRATEGIES FOR EFFECTIVE COLLABORATION?

Teams often go through stages to find what works best for them (Tuckman). However, understanding the differences between an effective vs. ineffective team as discussed in the previous section can serve as a guide to develop steps for planning and deploying your collaborative process.

STEP 1: ESTABLISHING COMMON GROUND AND VERBALIZING INDIVIDUAL EXPECTATIONS

Making time to get to know everyone and establishing common goals should be the first step of your collaboration. Whether or not this is required by your instructor, think of this step as an opportunity to establish common ground with your team members. This is also the time when team members should talk honestly and openly about their expectations and possible worries.

Setting expectations and creating a culture where everyone feels comfortable sharing ideas and setting their own learning goals will make your collaboration not just more enjoyable but also more effective.

Scholar Joanna Wolfe describes the importance of these preliminary conversations and provides a list of questions teams should address before starting their work together. The general spirit of these questions is to help team members to establish common ground by having honest conversations about everyone's goals and attitude toward the project. Here are some examples of questions based on Wolfe's approach that you may consider:

- What do you want to learn and get from this project?
- What level of effort are you willing to commit to this project?
- How interested are you in this project?
- What grade are you aiming for?
- What are specific concerns you have right now?
- What are your areas of expertise? What do you feel most comfortable doing?

Ideally, this conversation is held internally among team members, and not shared with your instructor.

STEP 2: ASSIGNING ROLES AND CHOOSING YOUR COLLABORATION METHOD

Depending on how your instructor sets up the team, this step may be completed prior to forming the teams. The goal is essentially to define how you will distribute the work. The method used, of course, should be appropriate to the project. In this regard, this step is also the opportunity to discuss and analyze together the main goals and requirements of the project. In other words, think of this step as the first close reading of the project's guidelines as a group.

What follows are some roles you may want to consider for a writing project:

- The project manager (sets the schedule; centralizes the communication within the team and between the team and the supervisor/instructor)

and, depending on the project,

- The researcher
- The content developer/writer
- The editor
- The designer

While the roles concerning the technical skills can be more or less flexible (ex. the editor can also be a researcher during the first phase of the project), the project manager should remain the same throughout the project. Assigning roles will not only allow you to use the team members' expertise to your advantage, but it will also make it easier to divide and distribute the work in a way that maximizes everyone's skills.

Dividing the work into different tasks with specific roles for each team member is the first step toward:

- Understanding the different components of the projects. Example: a technical report involves research; content development; design; editing.
- Making the project more manageable since team members can focus on their own areas of expertise and rely on their team members' contributions and expertise.
- Keeping the project cohesive, with input from all team members throughout the process.

Such an approach is sometimes called "a layered approach" because it involves team members taking turns in working on the final deliverable, adding their expertise at different stages of the process (Wolfe 6–8). "Norming" is another way to think about this approach, since each component of the project gets reviewed and worked on by all team members at different stages of the process (Richard Johnson-Sheenan 61–62). This is also a good reminder that the writing is never just done by one person; everyone does it at some point during the project. Assigning roles based on the project's goals and individual expertise keeps your work also more consistent with the realities of writing in the workplace.

Step 3: Identifying the Right Tools

Sometimes the biggest hurdle for a sound collaboration is to find the right tools. Fortunately, today's technology offers many options to work together without having to share physical artifacts or having to be physically together all the time. Regardless of whether the project involves word-based writing, visual design, or web development content, here are a few key questions to consider before choosing a specific tool:

- What are the collaborative features?

  Many online word-processors or content management systems have sharing options that enable team members to work together online, whether it is by adding content, editing, or commenting.

- How easy/difficult is it to archive everyone's contributions?

  A feature "showing the document's history" is interesting not only to document everyone's contributions but it is also a nice safeguard against accidental deletion or changes during the editing process. Think of this feature as a way to save your work as well as to document your input throughout the process. Having evidence of the work that you did, even if some of it may not end up as is in the final version of the project, not only helps to feel a sense of accomplishment but it may also work as a motivating factor and deterrent against freeloading.

- Does the tool guarantee that all team members have access to all the drafts?

    Saving drafts only locally is generally a bad idea. You want to make sure that your team sets up a shared folder where all the work related to the group project gets saved. Doing so will ensure that team members have equal access to all the drafts. Just like for the archive feature, this helps to give everyone a sense of ownership over the project by allowing members to intervene and comment as the project develops. Think of this element as the technical support for building and sustaining sound communication throughout the project.

When you discuss tool options with your peers, remember that the tools will not only help you to accomplish the tasks required by the project, but they will also allow everyone to participate and stay connected. In other words, consider the tools as another element for establishing common ground, building trust, and keeping everyone involved.

Step 4: Establishing Communication and Meeting Protocols

While it may be tempting to say, "we will decide on meetings as we go and as the project develops," it is paramount to establish a clear communication and meeting schedule at the onset of your project. Think of this as your communication framework for deploying your actual collaboration once you selected a tool (cf. earlier step 3). The goal is not to create more work but to facilitate your collaboration with regular check-ins. Here are a few elements to consider:

- What is the team's preferred communication method?

    Is it email, phone, text messages, instant messaging, social media messaging apps, or face-to-face meetings? You may also use a combination of methods. For instance, a team could use instant messaging for staying connected and reaching out when needed while also scheduling weekly face-to-face check-ins.

- How often will you check in with each other?

    Depending on the course and its modality, your instructor may give you time at the beginning of each class period to check in with your team. Regardless of whether or not you are given time in class, it is always a good idea to set your own check-in schedule. For example, for a month-long project, you probably want to have weekly check-ins. These check-ins may be time to share project updates, make sure everyone is (still) on the same page, and discuss next steps. They may also be time for seeking input from your peers on your contributions, asking questions, and solving eventual issues as a group (ex. The information you were supposed to gather is not available; your schedule changed, and you did not contribute as much as you had hoped this particular week). Check-ins not only streamline the work, but they also keep the stress level low since each team member will always know where the project is at.

STEP 5: WRITING THE ACTUAL GROUP CONTRACT/CHARTER

This step is the outcome of steps 1–4. Think of this contract as the first collaborative content that your team will develop and share with your instructor. As such, it should reflect everyone's input and vision. While the exact content may vary depending on the project, or how groups were formed, what follows are some basic elements that a contract should address:

- A summary of the team's main goals and the individual goals of each team member.
- A list of everyone's areas of expertise and their role/position.
- A description of your collaboration and communication methods.
- A list of the tools/technology you will use for the different components of the project (including for communicating, scheduling meetings, etc.).
- A project schedule with specific deadlines and outcomes for each component of the project. This also includes a schedule for reviewing each other's contributions and revising. (Refer here to your discussion about your communication protocols from step 4).
- Your teamwork policies. This should include a clear description of: the basic expectations; your approach to conflict resolutions and breach of contract (define together what this would be).

Once you have a detailed team contract, you are ready to start the work. Look at your contract as a map for initiating, planning, and deploying your collaborative process.

Table 6.1. Examples of Post-Collaboration Reflections Based on Whether or Not Each Step Was Followed.

	Did not follow the step	Followed the step
Step 1	I feel I did more than my peers. It was hard to ask questions. We worked but never really talked. During the whole project I felt I was just trying to go through the motions. I am glad it is done.	I was never afraid to ask questions or say something when I did not agree with a particular decision. We became friends, even though we had never worked together before.
Step 2	I never knew who did what or what I had to do exactly. It was a constant guessing game. At the end, we just assembled our different sections. The whole exercise just felt disjointed.	I always knew what my responsibilities were. Each of us contributed based on our strengths. Making time to edit and discuss what we had contributed really helped.
Step 3	One of us did the design and kept all the files stored on their computer because they had access to Adobe InDesign and that's what they wanted to use. So when they couldn't make it to class or to a meeting, we had no way of accessing the drafts. Plus, I couldn't edit because I do not have the program. This became super stressful in the end.	Having a common online folder with all the files for the project was very helpful. We could access the different drafts at all time. Working in an online setting that allowed us to comment on each other's work saved us so much time. Plus, I could work at my pace before class. We'd leave comments during the week on the draft and then take turns editing.
Step 4	We basically only interacted in class. While it was OK, we often wasted time trying to figure out who did what during the week. Not knowing if my team members were going to be in class or what they had done that week made the whole experience more stressful than it needed to be. We could have done a better job managing our time.	Deciding early on how we would communicate made it much easier to let each other know when something was not working or needed to be changed. We ended up meeting only once outside of class but we communicated almost every other day via text messages and in our online doc draft.
Step 5	We wrote a contract but never really followed it once we started. Looking back, I wish we had taken more time setting up how we would work, including how we would handle absences and such. Not having good policies made it difficult to have conversations about workload and absences.	There were a few times when some of us dropped the ball, but we knew each other well enough then to know how to handle it based on the policies we had agreed on. Having a plan in place helped us to adjust our roles when needed.

What does a Collaborative Writing Project Look Like in Action? The Collaborative Process Visualized

Figure 6.1 shows an example of what a collaborative writing project that uses the methods outlined previously looks like in action. Note that your team contract would be effectively written during the 3rd stage as the first content document after stage 1 and 2.

Common Problems and Possible Solutions

The methods outlined earlier offer a framework to build sound collaborative practices. They are, however, not full proof shields against every issue. What follows are some potential issues and solutions to keep in mind.

Social Loafing/Freeloading

If you end up with a non-contributing member, start by addressing the issue during one of the check-in moments. Give the person a chance to explain the situation first. Incomplete or no work is not always a sign of a freeloader.

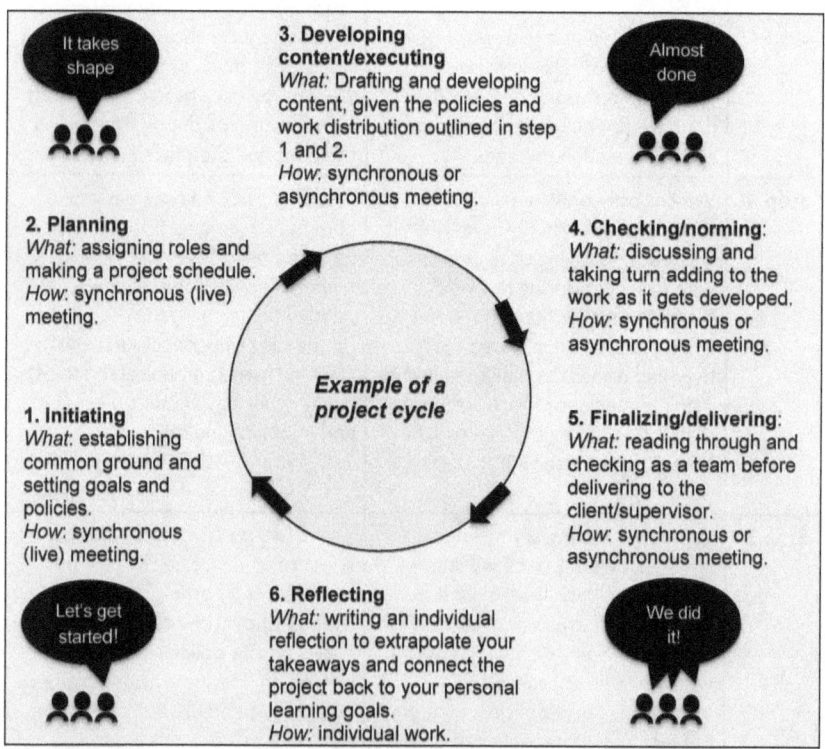

Figure 6.1: The Collaborative Process Visualized

Sometimes, a person does not contribute because their schedule or other unexpected factors get in the way. Sometimes they may also have difficulties finding the information for completing a task, or they do not fully understand what is expected from them. Having a conversation as a group will give you a chance to identify the source of the issue, provide support, and make necessary adjustments.

What started as a problem can then become a chance to build cohesion and trust. This may be enough to make a struggling team member a productive member. If these adjustments do not resolve the issue, consult your contract, which should have specific information regarding what constitutes a breach of contract and what procedure to follow. This may mean issuing a warning (with potential point deductions) and/or contacting your instructor. It may be helpful to organize a team meeting with your instructor as a facilitator. Remember to keep the focus on the problematic behavior and its consequences for the project, rather than on a particular individual. Make the outcome of the conversation what is needed to benefit the group and the project.

Unresponsive Team Members

To deal with members who go radio silent, apply the same steps as you would for non-contributing members. That is, start by giving a chance to the person to reconnect. Use the team's preferred communication method and keep in mind that their silence may be due to circumstances outside of their control. Someone may also withdraw because they are struggling with the work and need encouragement. If the silence persists, consult your contract (and schedule) and reach out to your instructor.

Conflict Between Team Members

Sometime disagreements or domineering personalities can create conflicts. If this happens, use your check-ins to have a conversation as a group. Give everyone a chance to speak without interruption and respect each other's ideas. Remember that the goal is not to be individually right or wrong; the goal is to be successful as a team. That is, understand that succeeding as a team may mean negotiating and compromising. Your project may include a peer assessment as one of the deliverables to be submitted at the end of the project. Such documents are helpful too for reflecting on collaboration issues and drawing lessons. However, if they are used as replacements for punctual assessments and check-ins during the project, group cohesion may suffer, and problems encountered during the project may become a major source of resentment and stress. In other words, do not delay and try to address issues as they arise.

Closing Thoughts

Collaborative writing projects are hard, as are any group projects, really. It is easy to focus only on the negative and the challenges and forget the benefits. However, as this chapter shows, there is much to gain from collaboration, including opportunities to develop relevant skills for the workplace, to get to know your peers better and make new friends, and to make your classwork more meaningful. In technical writing, group projects also offer unique opportunities to develop the kinds of skills required by today's professionals in the field.

Although there is no magic recipe for a successful collaboration, there are however specific strategies that one can use to plan and develop an effective group project. The key elements to remember is that collaboration is a process that takes time, effort, genuine engagement, and dialogue. In closing and just in case you may still be tempted to avoid collaboration at all costs or dislike the process no matter what, let's reflect on the words of scholars Cathy Davidson and Christina Katopodis: "Ditching group projects now, as the world increasingly depends on them, does not prepare our students for life beyond college." So, let's keep working together on opening the classroom door and making your writing count.

Works Cited

Allen, Nancy, and Steven T. Benninghoff. "TPC Program Snapshots: Developing Curricula and Addressing Challenges." *Technical Communication Quarterly*, vol. 13, no. 2, 2004, pp. 157–185, https://doi.org/10.1207/s15427625tcq1302_3.

Davidson, Cathy N., and Christina Katopodis. "8 Ways to Improve Group Work Online." *Inside Higher Ed*, 28 Oct. 2020, https://tinyurl.com/bdxyv7fc.

Duhigg, Charles. "What Google Learned From Its Quest to Build the Perfect Team." *New York Times*, 25 Feb. 2016, https://www.nytimes.com/2016/02/28/magazine/what-google-learned-from-its-quest-to-build-the-perfect-team.html.

Duin, Ann Hill, et al. "The Rhetoric, Science, and Technology of 21st Century Collaboration." *Effective Teaching of Technical Communication: Theory, Practice, and Application*, 2020, pp. 169–192, https://doi.org/10.37514/tpc-b.2020.1121.2.09.

Edmondson, Amy. "Psychological Safety and Learning Behavior in Work Teams." *Administrative Science Quarterly*, vol. 44, no. 2, 1999, pp. 350–383. *JSTOR*, www.jstor.org/stable/2666999.

Grant, Adam M. "The significance of task significance: Job performance effects, relational mechanisms, and boundary conditions." *The Journal of applied psychology*, vol. 93, no. 1, 2008, pp. 108–24.

Hall, David, & Buzwell, Simone. "The problem of free-riding in group projects: Looking beyond social loafing as reason for non-contribution." *Active Learning in Higher Education*, 14, 2012, 37–49, https://doi.org/10.1177/1469787412467123.

Johnson-Sheehan, Richard. *Technical Communication Today*. 6th edition, Pearson, 2018.

Murphy, Susan M., et al. "Understanding Social Loafing: The Role of Justice Perceptions and Exchange Relationships." *Human Relations*, vol. 56, no. 1, Jan. 2003, pp. 61–84, https://doi.org/10.1177/0018726703056001450.

Lane, Liz. "Interstitial Design Processes: How Design Thinking and Social Design Processes Bridge Theory and Practice." *Effective Teaching of Technical Communication: Theory, Practice, and Application*, edited by Michael J. Klein, The WAC Clearinghouse and UP of Colorado, 2021, 29–43. https://doi.org/10.37514/TPC-B.2021.1121.2.02.

Lam, Chris. "The Role of Communication and Cohesion in Reducing Social Loafing in Group Projects." *Business and Professional Communication Quarterly*, vol. 78, no. 4, 2015, pp. 454–475.

Pope-Ruark, Rebecca, et al. "Student and Faculty Perspectives on Motivation to Collaborate in a Service-Learning Course." *Business and Professional Communication Quarterly*, vol. 77, no. 2, 2014, pp. 129–149.

Tuckman, Bruce. "Developmental sequence in small groups." *Psychological Bulletin*, 63(6), 1965, pp.384–399.

"Identify dynamics of effective teams." *Re: Work With Google*. https://rework.withgoogle.com/guides/understanding-team-effectiveness/steps/identify-dynamics-of-effective-teams.

Williams, David L., et al. "Team Projects: Achieving Their Full Potential." *Journal of Marketing Education*, vol. 13, no. 2, 1991, pp. 45–53, https://doi.org/10.1177/027347539101300208.

Wolfe, Joanna. Team Writing: A Guide to Working in Groups. Bedford/St. Martin's, 2010.

Woolever, Kristin R. *Writing for the Technical Professions*. Pearson/Longman, 2005.

Teachers Resources

Overview and Teaching Strategies

Student collaboration does not have to be limited to full-fledged group projects. It can also be implemented at different stages of an individual project in a writing course:

- During the invention process of an individual project: students can work in groups to brainstorm and pitch their individual ideas to each other. Exchanging and testing their ideas among each other will make the benefits of collaboration more tangible.

- During the drafting and editing phase: creating peer review groups for students to assess each other's drafts is another way of taking advantage of collaboration for improving a written piece.

This essay can be taught to discuss and illustrate the benefits of collaboration for any stage of the writing process. By giving students ways to connect the WHY of collaboration to outside contexts and outlining specific strategies, the chapter can be used as a means to introduce a group project and/or to discuss at a more general level the value of peer learning in the writing classroom.

Discussion Questions

Though collaboration is used in many classes, it often is only introduced as a method that students are supposed to master without really being taught HOW to collaborate. Therefore, it is essential that teachers address the HOW and the WHY of collaboration before asking students to develop any project in a group.

Here are some preliminary questions to introduce and contextualize collaboration as a learning topic and outcome:

1. What experiences do you have with collaboration?
2. What are the biggest obstacles to and disadvantages of collaboration?
3. What are the main benefits of collaboration? Why do you think group projects are assigned? What makes group projects particularly relevant in the technical writing classroom?
4. How would you define effective collaboration?

Such an introductory discussion encourages students to share their experiences (bad or good) and makes collaboration not just a method but a topic and tangible learning outcome.

This discussion can be supplemented with a short article on collaboration in the workplace and what makes teams effective (ex. The Google Aristotle Project). Doing so will help emphasize that sound collaboration does not just happen but needs to be taught and developed over time. Another activity that can be combined with this introductory discussion is asking students to do an online search on "collaborative skills" and describe what employers and recruiters say about it in their respective fields.

Questions pertaining to specific collaborative strategies could include:

1. This chapter describes specific strategies for implementing a successful collaboration. What are some examples?

2. What is the goal of a group contract? What should it contain?
3. How can a group distribute the workload so that it is efficient and remains collaborative?
4. What are some communication strategies a team can use to resolve conflicts?

By inviting students to reflect on collaborative practices (including the importance of communication), the chapter gives students ways to plan a project and shows them that there are specific strategies for making collaboration effective.

Technology Activity/Discussion

This activity can work as a complement to an introductory discussion on collaborative writing projects. It can also be assigned as an exercise requiring students to analyze specific writing technologies and identify their effectiveness in collaborative settings.

1. What software programs have you used for collaborative projects? List their pros and cons.
2. What are the differences between synchronous and asynchronous collaborative mode? Which technology can you use for each mode?
3. What kind of software programs does the project require? List a few options. Test and rank them based on their collaborative options.

Responding to Collaboration Planning Exercise

Scenario-based exercises can also be used as application of this essay. Here is an example: a group of students was assigned a project requiring them to develop instructions for an online application of their choice. After reading the project guidelines individually, they were given time to plan their collaboration. Consider the following exchange:

> **Student A.** "It looks like we have a lot of leeway to choose the application we want to use. I have an idea that I can make work easily. I will start writing something and we can look at it during our next class. What do you think?"
>
> **Student B:** "This sounds good. I will look up images that we can use. I like doing that kind of stuff anyway."
>
> **Student C:** "Cool! Let me know when you have something for me to look at. I'm happy to edit. Just let me know how I can help."

How does this initial exchange work or does not work for planning the project? What are the pros and cons of such an approach? Discuss these questions with two or three peers and share the feedback you would give to this group with the rest of the class.

7 Worth a Thousand Words: Constructing Visual Arguments in Technical Communication

Candice A. Welhausen

In June of 2018, the cover of *National Geographic Magazine* called attention to a severe and rapidly growing environmental problem.[1] Using the organization's trademark yellow border and logo, the issue showed an upside-down, white, plastic bag—like you would get at a grocery store—mostly submerged in clear blue water. In this image, the left corner of the bag rises above the horizon line, twisting to form a peak, while the handles of the bag float below, sinking into the depths of the darker blue water. The title of the issue, "Planet or Plastic?," along with a short descriptive caption to the right of the floating bag— "18 billion pounds of plastic end up in the ocean each year. And that's just the tip of the iceberg"—reinforces the overall visual metaphor constructed through this imagery. In other words, plastic pollution is a dire problem. It's getting worse. And we need to address it immediately.[2]

As this example demonstrates, text and visuals often work together persuasively to make arguments and construct information in particular ways. Indeed, the visual and linguistic (text-based) *rhetorical* or argumentative choices that the creators of this cover made—using the white, plastic grocery bag; manipulating it to look like an iceberg; positioning it at eye-level; using a light to dark blue color gradient for the water; writing the caption—are all highly intentional. Because any number of images could have been used to show the plastic pollution problem—an empty beach strewn with trash, a

---

1 Due to copyright restrictions, this image cannot be reproduced here. But it can be found on National Geographic's website at https://nationalgeographicbackissues.com/product/national-geographic-june-2018/.

2 This work is licensed under the Creative Commons Attribution-NonCommercial-NoDerivatives 4.0 International License (CC BY-NC-ND 4.0) and is subject to the Writing Spaces Terms of Use. To view a copy of this license, visit http://creativecommons.org/licenses/by-nc-nd/4.0/, email info@creativecommons.org, or send a letter to Creative Commons, PO Box 1866, Mountain View, CA 94042, USA. To view the Writing Spaces Terms of Use, visit http://writingspaces.org/terms-of-use.

bulldozer burying debris at a landfill, a marine animal caught in a fishing net. But the creators of this cover chose a white grocery bag probably because they thought that most viewers would instantly be familiar with this type of plastic. All plastics used for consumer products—shampoo bottles, take-out containers, food packaging—contribute to plastic pollution. That said, a great deal of attention has been focused specifically on these bags because it's easy to use a lot of them during a single trip to the grocery store. And many people just throw them in the trash after unpacking their groceries.

The creators of this cover may have also decided that "tip of the iceberg"—a common metaphor used in language—could easily be visually extended to these bags because many of them are white. They probably also thought about the blue shading they would use to symbolize both the depth of the ocean and the depth of or extent of the problem, which would also visually emphasize the white color of the bag. In carefully constructing their overall visual message, they may have also thought that lightening the twisting top of the bag and darkening the blue below the bag's sinking handles would help visually reinforce their overall visual/verbal argument.

As a working professional, you may not be creating magazine covers like the one described here. But you will likely find yourself in situations where you need to create other types of visual communication. For example, many scientists and engineers construct technical drawings like illustrations and sketches. Professionals working in business and marketing may need to visualize financial data by creating tables and figures. And communications experts might use photographs to create brochures, pamphlets, and other educational materials. In turn, the audiences that you create these graphics for—supervisors and managers, co-workers and peers, customers, and clients—will then use these visuals to make particular kinds of decisions and/or take particular kinds of actions.

This chapter discusses these visuals and the ways they might be used persuasively. It also provides strategies that can help you create specific visual forms. The chapter begins by describing the semiotic functionalities of visuals—that is, the way that visuals convey particular meanings. The chapter then discusses several visual forms of communication—photographs, technical drawings, and data visualizations—in more detail. It concludes with an extended example that describes how visuals were used in two documents—a brochure and a calendar—created by students enrolled in a graduate technical and professional communication course for Rural Studio, a program in Auburn University's School of Architecture, Planning and Landscape Architecture. This "off-campus design/build program," as Natalie Butts-Ball, the communications manager for the program, explained, was established in 1993. And it allows architecture students to gain hands-on experience by assisting "under resourced rural communities in West Alabama," she added.[3]

Guiding Questions: As we begin examining visual forms consider the following:

- What kinds of visual communication are common in your field?
- Who are the audiences for these visuals?
- How do these audiences use these visuals? In other words, what decisions do viewers need to make and/or what actions do they need to take **after** viewing these forms?

By considering the answers to these questions, you can create more effective visuals for the technical and professional documents you create.

Semiotics: The Study of Signs in Visual Communication

Signs and Signifiers

Envision that it's the first week of the fall semester. You're walking across campus to meet some friends on the lawn in front of the student center. As you approach the building, you see two of your friends—Kaitlyn and Madison. Kaitlyn runs over to Madison, arms outstretched. Madison gasps and raises her hands to her face. She then throws up her arms, and the two women hug each other.

From afar, you couldn't hear their conversation. But you knew they had not seen each other in a while (and were happy to be reunited) because you *read* their body language. Such non-verbal communication cues are key to the ways that humans communicate with each other. We also use other semiotic modes or *sign systems* like language (both spoken and written) and visuals.

Generally speaking, a sign—words, gestures, images—is something that stands for something else. Because we can't read each other's minds, we need to use signs to communicate. The word *chair*, for instance, is a linguistic sign that refers to a piece of furniture with four legs and a back—like the image in Figure 7.1.

---

3   The author thanks Matt Crouch who created the Breathing Wall brochure and Natalie Butts-Ball who created the Harvest Calendar for allowing me to include and discuss their design work in this chapter. Both can be foujnd in the appendix. The author also thanks the students of the Rural Studio Program who contributed to these documents.

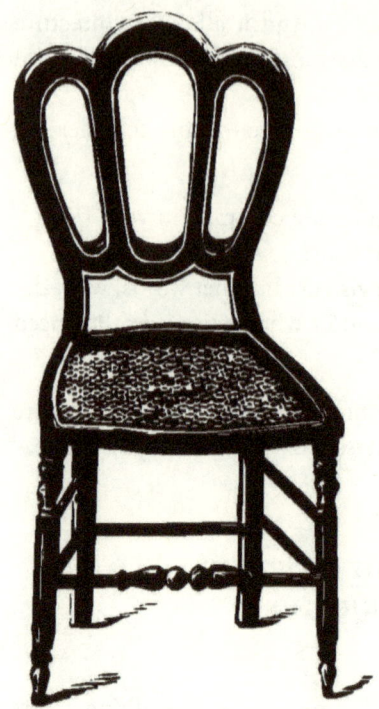

Figure 7.1: Chair

Swiss linguist Ferdinand de Saussure theorized that signs are comprised of a *signifier* (e.g., the sign) and a *signified* (the thing or concept the signifier refers to). More specifically, the word *chair* and the image in Figure 7.1 are both *signifiers*. These linguistic and visual signs refer to the concept of *chair* (the *signified*). The image of the chair in Figure 7.1 is more specific than the word *chair* because it shows a particular style of chair. However, if we were talking about chairs, we could also add an adjective or two to be more specific such as *office* chair or *red, lounge* chair or *old, vintage* chair like the visual sign in Figure 7.1.

Sign Systems

Collections of signs—groups of words, for example, —comprise sign systems. Indeed, language is a sign system. And sign systems have rules that determine how signs are used and interpreted—like grammar and punctuation. Visual communication researchers Gunther Kress and Theo van Leeuwen have proposed that visual communication, too, uses a "grammar." Yet unlike the grammatical rules for language, which you learned about in grade school, you probably have not given much thought to the *grammatical* system you use to interpret images. Kress and van Leeuwen have said that the meaning of images often seems intuitive because "we know the [semiotic] code already" (32). In other words, you probably don't have to think much about how to interpret visuals because you know what they mean.

To illustrate, color choice is understood primarily through emotional associations, as technical communication scholars Nicole Amare and Alan Manning have argued. Red, for instance, is frequently perceived as alarming—many warning labels use red text, a red background, and/or red imagery. Blue, on the other hand, is frequently perceived as calming. But red can also be used to convey excitement or to grab viewers' attention. And blue can be used to mean sadness. The meaning of different color choices, then, also depends on the context in which the color is seen as well as how it's used. There is nothing inherently alarming about red or sad about blue. But we perceive these colors as such because these associations have become established through repetition and cultural conventions—like seeing the same color repeatedly used to convey the same idea in the same situation. For example, because red has consistently been used to convey alarm, when we see this color on a safety information card, we know that it's being used to warn us. In other words, we already know what red *signifies* in this context.

DENOTATED AND CONNOTED MEANINGS

French theorist Roland Barthes lends further insight into signs by proposing that we interpret images through layered meanings that are both *denotated* and *connoted*. The denoted meaning of an image refers to what is actually shown—like the chair in Figure 7.1. In other words, the denoted image is the literal semiotic message. But we do not always interpret signs solely by their literal meanings. As the discussion about color demonstrates, context shapes meaning as do the personal, cultural, and historical values that have become embedded in particular visual signs and combinations of signs.

To illustrate, the chair in Figure 7.1 is illustrative, included here to provide an example of a visual sign. However, if you saw this chair in a different context, in an advertisement that popped up as you were shopping online for a new desk chair for your dorm room or apartment, for instance, you would probably interpret it differently. Let's say the company selling the chair also sells outdoor furniture. In the advertisement, the chair sits next to a matching table in an outdoor setting. You're not looking for this type of furniture right now. But as you click past the ad, you're briefly reminded that a close friend of yours used to have a chair and matching table kind of like those shown. For a moment you momentarily feel nostalgic as you think about your friend and the last time you saw her before focusing your attention back on the task at-hand, finding the right chair for your desk.

Such associations are the *connoted* meanings of images. Meaning, when you viewed that advertisement (or any image for that matter), you understood the literal visual signs—chair, table, outdoor setting. But you also assigned meanings to those signs based on the values and emotions that the imagery may have invoked. An outdoor setting, for instance, —whether it seems familiar or not—can create particular kinds of associations for viewers, which are culturally engrained. For instance, an advertisement for soft drinks, for instance, that includes a group of people laughing and preparing food around an outdoor grill might remind American viewers of spending time with family and friends during summer holidays—like a Fourth of July barbecue, for instance. Such events are common and highly valued by many people from this culture. Viewers from another country, however, might have very different associations that come to mind when viewing the same imagery.

Going back to the advertisement for outdoor furniture, envision that there are also other visual signs like a laptop sitting on the table, for instance. Viewers will also automatically assign particular cultural values to this visual sign. For viewers from some cultures, including a laptop in this outdoor setting might signify a relaxed and flexible work environment. But for viewers from cultures who don't share this value, the laptop might be perceived as inappropriate if it's too informal or if people from that culture don't tend to work on laptops in that kind of environment. In this example, the laptop connects the cultural associations of the outdoor setting to workplace practices, invoking emotions that are linked to those associations. These connections construct what Charles A. Hill described as a "three-way relationship . . . between the image, the value . . ., and the emotions that are . . .linked to that value" (35).

Understanding and Interpreting Signs

The fictional advertisement for chairs described in the previous section can be understood both denotively and connotatively. However, some visual signs are interpreted entirely symbolically because they represent abstract ideas rather than concrete things. For instance, for many individuals in the US, an image of the American flag is very rarely interpreted literally. Rather, it's understood in terms of the cultural values it signifies: independence, individual freedom, equality.

When the flag is used in imagery, these values become embedded in the overall semiotic message. For instance, in the introduction to a book on visual communication, Marguerite Helmers and Charles A. Hill argue that the American flag is central to the way that the photograph, "Raising the Flag at Ground Zero," was interpreted. Taken the day following the 9/11 terrorist attacks in 2001, the image captures three firefighters hoisting an American flag against the backdrop of the destroyed World Trade Center site in New York City. These men raised the flag, Helmers and Hill explain, "to rally the spirits of those working amidst the rubble of the Trade Center" (6). Indeed, the flag is the focal point of the image. But it's also the primary visual semiotic message, conveying "drama, spirit, and courage in the face of disaster," as Thomas Franklin, the photographer, described it (qtd in Helmers and Hill 7). Such iconic photographs, as discussed in more detail in the next section, can come to represent particular historical moments.

A key point to remember is no object, item, or image inherently has any meaning. Rather, we **learn** to associate meaning with particular things over time—either by being taught a given meaning (e.g., a teacher telling us the color purple is associated with royalty) or through repeated observation (e.g., always seeing the color green used to signify *go* in traffic lights). We also often associate this meaning with the specific context in which we see things—green means *go* when we see a traffic light, but it means *environmentally sustainable* (e.g., *green* products) when we see it used in the packaging for a cleaning product at the store, for instance.

Guiding Questions: At this point, you may be wondering: what does semiotics have to do with the visual information I will be creating? Everything. Regardless of the specific visual form you construct—a photograph, technical drawing, or data visualization—you will be *signifying* particular ideas through the decisions you make. Consider the visual signs that you want to use in a particular communication situation—that is, the imagery, the color choice(s), and the subject matter of your visual content. Then consider how your viewers are likely to interpret these signs within that communicative context.

- What are the denotative and connotative meanings?
- Are these the meanings you want to convey?
- What vales do your choices reflect? Are these the values that you want to convey? Will these values resonate with your viewers?
- Are there other visual choices that might be more effective and persuasive for your purposes?

The core item to keep in mind as you think through these questions is to make these decisions based on the assumptions you have made and/or the research that you have conducted on your audience. If you are creating visual information for viewers from your same cultural background, you can probably assume that they will interpret many signs similarly like color, for instance. However, if you are designing information for intercultural or international readers and/or readers from a culture that is different from your own, it's important to research your audience before making your visual choices.

Photographs

A Brief History

Photography has a long history of being perceived as showing the world as it really exists. Invented in the nineteenth century, it has since become ubiquitous in everyday life. In fact, most of us do not give a second thought to grabbing our smartphones and snapping a picture when we want to capture something that interests us. We might then upload the image to social media, and save it for informational purposes or to show to other people.

As a form of visual communication, photographs and video are among the most "vivid" forms of information because we perceive them as closely approximating "actual experience" (31), scholar Charles A. Hill argued. However and despite photography's longstanding "legacy of objectivity," as Sturken and Cartwright (28) put it in their book on visual culture, photographs have never been neutral. When you take a picture with your phone, for example, you make a decision about the kinds of experiences and/or information that you think is worth documenting. And all of the choices you make when you take the picture—whether you move closer to your subject or step back, use filters and/or special effects, ask your subject to look at the camera or spontaneously capture the scene—shape the image you ultimately create.

During the late twentieth century, photography evolved into its current digital form through tools like photo editing software, which allows us to easily manipulate these images. Most recently, the emergence of increasingly sophisticated AI tools such as those that let users create highly realistic *deepfakes*[4] as well as *generative* applications are increasingly blurring the line between fact and fiction. These technologies are easy to use and users can quickly create a wide variety of images. But these tools can also perpetuate dangerous sterotypes and discriminatory practices as Joy Buolamwini discovered when she was a college student at MIT working on facial

---

4 This term is a combination of "deep learning" and "fake." It refers to an image or a video of one person that has been altered to look like a different person such as superimposing the head of one person onto the body of someone else.

recognition software. In an article she published describing her experiences, Buolamwini discussed two projects that she worked on wherein the AI technology could not recognize her dark skin. In one instance, she had to use the face of her white roommate and in another she had to use a white mask to finish her work. "A.I. systems," she advises, "are shaped by the priorities and prejudices — conscious and unconscious — of the people who design them" (Buolamwini, para. 3). As this example illustrates, we need to pay attention to and be highly critical of the content being produced by **all** AI technologies—both visuals as well as the written content that chatbots are increasingly being used to create. Because these tools embed and normalize the values and biases of the people who created them in ways that may not always be immediately obvious.

SUBJECT MATTER AND CAMERA ANGLE

Regardless of the technology used, photographs construct a particular version of reality. Consider the well-circulated image entitled "Taking a Stand in Baton Rouge" taken by Jonathan Bachman for Reuters on July 9, 2016 during a protest in that city against police brutality.[5] In the photograph, a Black woman, Leshia Evans, stands in front of two police officers who walk toward her to take her into custody for blocking a highway. Years later—in the late spring and throughout the summer of 2020—images showing confrontations between protesters and police would become commonplace not just in mainstream news coverage but on social media as civilians documented these encounters. Yet this particular image is remarkable, in part, because while more recent photographs often show these interactions from a distance, this image captures the encounter from the perspective of an immediate bystander. More specifically, Bachman's eye-level camera angle allows us to experience Evans' encounter with the two officers as though it were happening right next to us.

In the full-length shot, Evans and the officers are centered in the frame, which visually reinforces the centrality of the three people involved. A small crowd of onlookers can be seen in the background to the right, while a line of police officers flanks the background to the left. A news article about the photo describes Evans' demeanor as "calm," the thin fabric of her long summer dress billowing lightly around her calves as she faces the officers directly.[6] Her stance, the article explains, as a commenter on Face-

---

5   Due to copyright restrictions, this image cannot be reproduced here, but it can be found at this location: https://www.bbc.com/news/36766799

6   See https://www.bbc.com/news/world-us-canada-36759711

book put it, is "balanced, powerful, upright and well grounded . . . She is only protected by the force of her own personal power. By [sic] contrast, the officers have the transitory, temporary, protection of their equipment that will be removed at the end of their shift."[7]

Bachman stated that he felt "very humble to capture an image that tells the story of what has been happening here in Baton Rouge."[8] Indeed, he captures the experience of the three people shown at a particular moment during the protest as Evans and the officers also likely had other experiences there. For instance, Evans may have been part of a larger group at some point, and the two officers shown may have had interactions with other protesters. Many of the other protesters as well as other police, onlookers, and passers-by who were there that day are also not captured in the image. Further and in all likelihood, not all of the protestors who where there were directly confronted by the police. And not all of the police on duty may have interacted with protesters. Not all protesters who attended may have been Black and female and in their mid-30s—like Evans. And probably not all of the police officers who attended were white and male—like the two officers in this image. In other words, Bachman's image tells *a* story about the protest. But it's not the whole story, and it's certainly not the *only* story that could be told about what happened at this protest event in Baton Rouge that day.

Iconic Photographs

At the same time, this photo has also become iconic—representative of this event and arguably others like it—because it visually promotes a particular narrative laden with American cultural values: individualism, opposing injustice, non-violent protest—to name a few. Indeed, this image has been compared to another photograph taken nearly thirty years earlier, "Tank Man," which showed a lone man blocking a line of tanks during protests in 1989 against the Chinese government at Tiananmen Square.[9] The context in which this earlier photo was taken is different as is the subject matter. However, arguably the underlying message is the same. As a commenter on social media put it: Evans is "speaking truth to power," although she never says a word.[10]

---

7 See https://www.bbc.com/news/world-us-canada-36759711

8 See https://www.bbc.com/news/world-us-canada-36759711

9 Due to copyright restrictions, this image cannot be reproduced here, but it can be found at this location: https://en.wikipedia.org/wiki/Tank_Man

10 Due to copyright restrictions, this image cannot be reproduced here, but it can be found at this location: https://www.bbc.com/news/36766799

Figure 7.2: Migrant Mother (Dorothea Lange, 1936. Public domain.)

Iconic photographs, journalism professors Robert Hariman and John Lucaites have argued, need "no caption." Like other photographs, these images capture a moment in time. But more importantly, they come to represent the historical significance of that moment. For instance, you have probably seen "Migrant Mother" (see Figure 7.2), the now famous photograph captured by a documentary photographer during the Great Depression in the US.

In this image, the clothing of the mother and her children is worn and tattered, visual signs that signify poverty. The children's faces are hidden from the viewer, burrowing behind the woman. One child's folded arm hangs on the woman's shoulder. The woman leans forward, her right hand cupping the bottom of her face, indicating deep and troubled thought, visually signified by her furrowed brow. She faces the viewer, but her gaze fixates past us on something further away on the horizon. Something far more important. More urgent. The struggle she is experiencing at this moment? The economic hardships yet to come?

The choices that Lange, the photographer, made in composing this image—choosing to capture the woman and three of her children in this particular image rather than the entire family (including the woman's husband), capturing her subjects from the waist up (rather than from a distance, which might have included the family's tent in the background), framing them at eye-level (rather than below or above), waiting for just the right moment to click the shutter—all of these factors influence how we interpret this image. Indeed, making other compositional choices would have resulted in a different photograph, and Lange took other pictures of this migrant family. But the close angle here allows us as viewers to experience the full emotional impact of the woman's expression.

Migrant Mother has been so widely circulated that the visual signs in this photograph are no longer interpreted as showing one American family's desperation, poverty, and hopelessness. Rather, Migrant Mother documents the desperation, poverty, and hopelessness that *all* Americans experienced during the greatest economic collapse in US history and that we now associate with that event.

STRATEGIES FOR USING PHOTOGRAPHS PERSUASIVELY

As you think through the information in this section, carefully consider how the rhetorical choices discussed (e.g., subject matter, camera angle, perspective) will influence how the photographs you use in your work are perceived—that is, how the visual choices you make will create particular effects and interpretations.

To illustrate, envision you are a healthcare provider for an urgent care clinic, and you need to create a brochure explaining the clinic's services. You'll include text in your brochure, but you'll also want to *show* readers what they can expect during their visit—the kinds of procedures that will be done and the kinds of interactions they will have with clinic personnel. For instance, after patients check-in, a nurse will escort them to an exam room, and check their vital signs including their blood pressure.

You decide to include a photo of a patient having her blood pressure checked.

During the patient's visit, the healthcare provider will ask the patient questions about her health status and history. So, you also decide to show a provider talking to a patient in an exam room. Regardless of what imagery you decide to show, you'll want to ensure that the people in your photographs are similar in age, gender, and racial/ethnic background to the population that the clinic serves. If the clinic treats women and men, you will want to include photographs of both. But if you create the brochure for a women's health clinic, you'd likely only include images of women. Similarly, if the clinic primarily serves the elderly or children, the patients in your photographs should reflect these age groups.

In terms of camera angle, you'll want to consider how to best frame your subjects (e.g., providers and patients) to convey one-on-one interactions. So, you decide to show the provider from the patient's perspective, at eye-level and with the patient's back to the camera and the provider's face clearly visible. Both the patient and the provider are seated, and the provider is looking down at her notes, preparing to talk to the patient. You make these decisions so that patients can envision themselves from the perspective shown in the photograph—that is, sitting and talking to the provider. Overall, you want to show readers what providers might look like, how they might interact with patients, and the kinds of medical equipment they will probably use. By making these choices, you are making a visual argument about the kind of experience that patients can expect to have at the clinic. And patients will use your brochure to decide if they want to receive healthcare there.

Guiding Questions: When working with photographs, consider the following rhetorical choices:

- What subjects are you choosing for your images and why? Why did you choose a particular subject matter for a particular communicative situation? What are some alternative subject matters that might have conveyed the same idea? Why did you not choose these options?
- What camera angle are you using (e.g., low, high, eye-level, at a distance), and how does this choice influence how viewers will perceive the subject?
- What is the compositional perspective (e.g., close-up, full-length, at a distance), and how does this influence how viewers will perceive the subject?
- What is the decision that your viewers will need to make and/or the action they will need to take after viewing your photograph?

Technical Drawings

As the previous section demonstrates, when you take or use photographs, you are making rhetorical decisions about what to show (e.g., the subject matter you choose) and how to show it (e.g., camera angle and compositional perspective). Indeed, you may find that in some communicative situations you need and want to use highly realistic images like photographs. However, it might be more appropriate in other situations to use more abstract visuals.

Illustrations

In his work on comic books, Scott McCloud argues that abstraction allows viewers to fill in details and construct their own meanings more easily. For instance, healthcare providers often use illustrations—like the image of the human heart in Figure 7.3—to explain medical procedures to patients. Let's say you're a heart surgeon, and a patient needs surgery to repair a heart valve. You might use Figure 7.3 to explain how blood moves through the patient's heart. More specifically, areas on the left side in this drawing indicate the flow of oxygen-poor blood returning to the heart after being circulated throughout the body, whereas the darker and smaller chamber on the right side of the image indicates oxygen-rich blood that is ready to circulate.

In theory, you *could* use a photograph to explain this process. However, it would probably be too graphic for patients. And design elements like shading, contrast, lines, and color (not shown in Figure 7.3) can be used in illustrations, allowing viewers to more clearly see the different areas that the creator of the graphic wants to show—like the parts of an organ as Figure 7.3 demonstrates. Indeed, it would be difficult to see the different chambers of the heart in a photograph. And lines and arrows can be added to an illustration (like Figure 7.3) to show viewers the process of blood flow, for example, which would not be effective in a photograph.

Photographs can also be difficult to modify in an office setting where you may have a limited amount of time to explain a procedure. More specifically, as you explain the medical procedure to the patient, you might want to circle the value that needs to be repaired or add notes and/or details. In Figure 7.3 the differently shaded areas and the arrows showing the direction that blood is moving make it easier for you to explain to the patient how the heart functions because the components and processes of the heart are visually simplified.

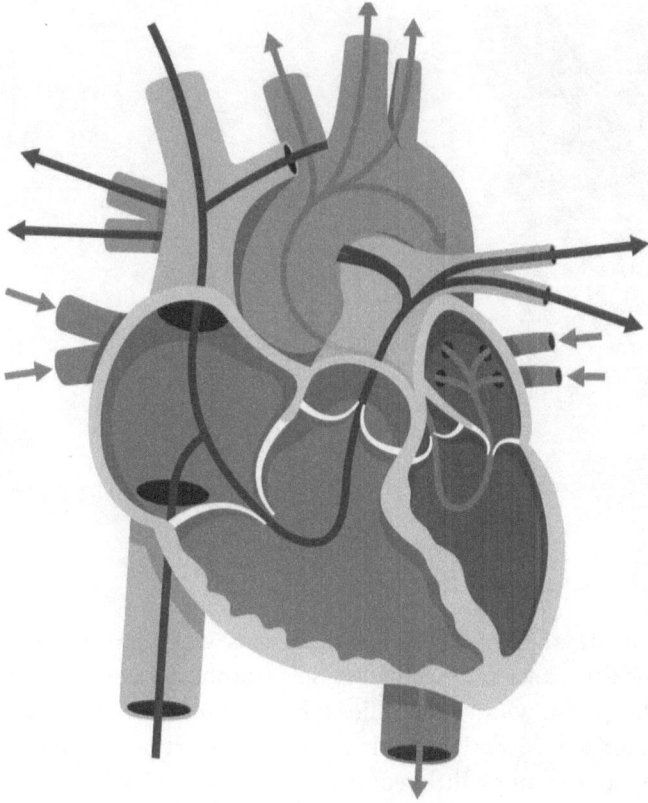

Figure 7.3: Illustration of the human heart

Sketches

Indeed, because they are less *real*, abstracted images like drawings, illustrations, and sketches are also more *ideal*. That is, they allow for more flexibility in the way they are interpreted, and they can be modified more easily. These images also often remove a lot of non-essential visual information from an image and allow individuals to focus *only* on those visual elements that they need to understand an idea. To illustrate, the sketch in Figure 7.4 shows a fictional schematic drawing for a bridge. The perspective grid lines that divide the image allow viewers to see how the structure will occupy space once it is built. And because this visual is a sketch, changes can continue to be made to the overall design (either digitally or on a printed copy). Indeed, the lack of contextual details—where the bridge will be built, how it might affect the landscape and other built structures—allow the creator to focus on the details of the design.

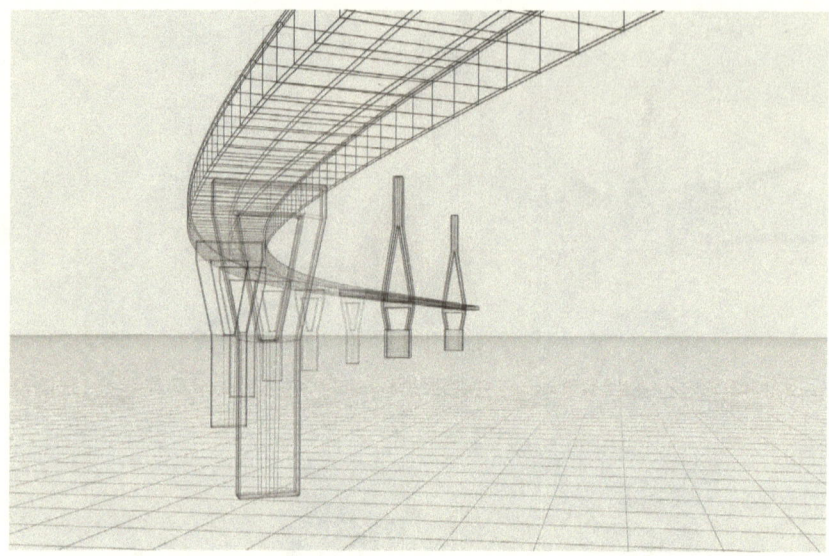

Figure 7.4: Schematic bridge diagram

Creating such visual forms allows professionals in design-oriented fields like engineering and architecture to solve problems and explore ideas. As Donald Schön, who was a philosopher and professor of urban planning characterized it, design work involves "reflection-in-action" wherein creators work through different design options. Indeed, a structural engineer might have created the sketch in Figure 7.4 to assess and make decisions about the height and width of the pylons—as well as other components of the structure—as she fleshed out the details in the design of this bridge.

STRATEGIES FOR USING TECHNICAL DRAWINGS PERSUASIVELY

As you think through the discussion in this section, carefully consider when you might want to use a more abstracted visual representation like a drawing, illustration, or a sketch rather than a photograph. Because this decision, too, will have a particular rhetorical effect.

To illustrate, envision you are an architect, and you have a client who wants to build a three-story apartment complex. You meet with the client, and she provides details about the proposed building including its location and dimensions, as well as her preferred architectural style. You use this information to brainstorm a few ideas using your design software, and you create a few draft schematic diagrams—like the image shown in Figure 7.4. You share these with the client, and she likes your last design the best.

So, you scrap the others and start adding details to flesh out a more concrete image. You create a full color, three-dimensional digital model of the entire building including landscaping features and floor plans. You set up a meeting to show your work to your client. The client will then use your visuals to decide if your designs align with her vision for the building. And you'll most likely continue to make changes as you and your client work on the design.

Guiding Questions: When you create drawings, illustrations, and/or sketches, carefully consider the following:

- Why are you creating this type of image rather than using a photograph, for instance? In other words, why do you need a more abstract visual rather than a graphic that is more realistic?
- How much detail do you want to include in the image? Should your visual be more abstract or should it be more fleshed out for your rhetorical purposes?
- What is the decision that your viewers will need to make and/or the action they will need to take after viewing your drawing, illustration, or sketch? How do your design decisions help viewers make this decision or take this action?

Data Visualizations

Finally, as a working professional, you might also need to visualize quantitative information by creating graphics like tables, figures, thematic maps, and/or infographics. Like the other visual forms discussed in this chapter, you will need to make decisions about what information to visualize and how to visualize it, depending on your intended audience and how they will need to use the information.

Bar Charts and Line Graphs

For instance, let's say you're a local sales manager for a company that manufactures several different kinds of widgets. The company's older models (A, B, and C) sold well last year. However, your boss—the regional sales manager—wants to increase sales this year for the newer models (D and E). You schedule a meeting with your marketing team to brainstorm ideas. In preparation, you create the bar graph shown in Figure 7.5 that includes total sales for each model over the past year. During the meeting, you and your team use this chart to discuss strategies to increase sales for models D and E.

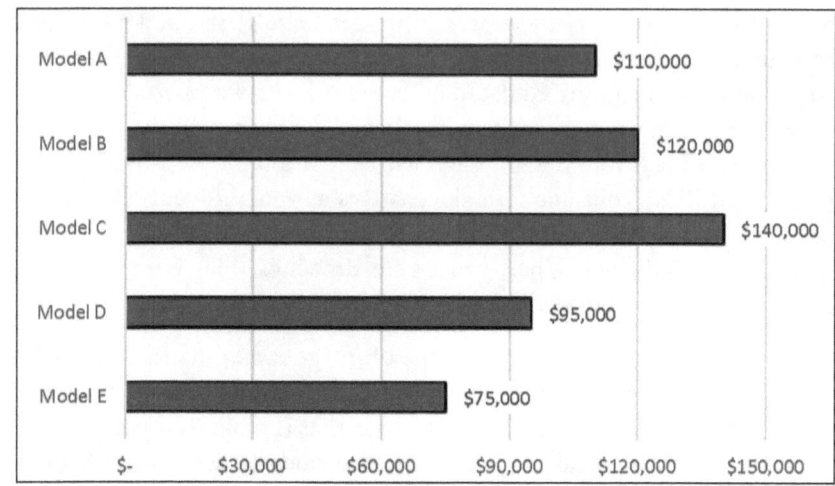

Figure 7.5: Yearly sales for each widget model

In this fictional scenario, you created a bar chart so you and your team could compare total sales by model for the year. Creating a different type of graphic—like a line graph—would be useful if your group wanted to see sales by month for one or more of these models as shown in Figure 7.6. While Figure 7.5 allowed your group to compare sales totals, Figure 7.6 allows you and your team to see trends in sales over the year by month. Figure 7.6 shows that sales for all models tend to be highest in the summer with a dramatic increase for all in July in particular. Sales then tend to drop off in August and continue to decline into the fall and winter.

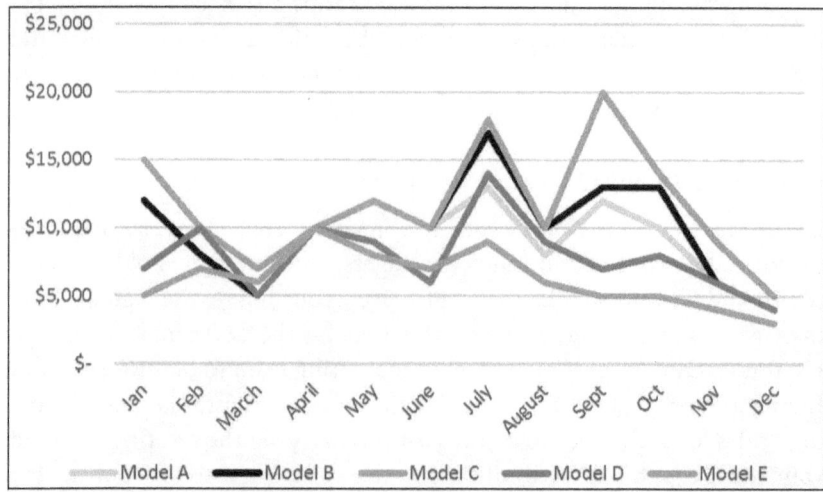

Figure 7.6: Sales by month by model for last year

Thematic Maps

If your team is also interested in where most of the sales were occurring in a particular region, you could have also created a thematic map. An example of this visual genre is shown in Figure 7.7. Overall, the information you visualized—the bar chart shown in Figure 7.5 and the line graph shown in Figure 7.6—enabled you and your team to see patterns in your company's sales data and to ultimately make decisions about marketing strategies you might want to use to increase sales for the underperforming widget models (D and E). At the same time, by creating these graphs, you also made a visual argument directed toward your team that the company's sales information should be interpreted in a particular way—that is, comparatively (i.e., by using a bar chart) and temporally (i.e., by using a line graphs) rather than spatially (i.e., by using a thematic map). In other words, you decided that the total dollar amount of sales (Figure 7.5) and total sales for each month (Figure 7.6) was important while the location of these sales was not. The type of graphic that you choose—and the specific design choices that you make in creating these graphics (e.g., color choice, line style, scale)—also shape and influence how quantitative information will be perceived and subsequently interpreted.

Infographics

This idea can be seen in two infographics (see Figure 7.7) that were published in July of 2020 during the COVID-19 pandemic. More specifically, the Georgia Department of Public Health came under fire for publishing what some argued were misleading thematic maps when Andisheh Nouraee, an Atlanta-based communications and marketing expert, tweeted screenshots of the infographics shown in Figure 7.7.[11] His tweet, which quickly went viral, pointed out that the number of COVID-19 cases by county in the state increased by 49% between July 2 (left) and July 17 (right). However, visually, we do not perceive much (if any) change in the shading used in the maps because as Nouraee points out, the scale of the legend has changed (see note 11 below for a link to a color version of these maps). On July 2, red (not shown) was used to show counties with between 2,961 to 4,661 cases. But on July 17, red (not shown) was used for counties with 3,769 to 5,165 cases—a 49% increase, he notes. These changes are not visually evident when comparing these maps.

---

11 Nouraee's tweet is no longer available on Twitter. A copy of the maps, and a discussion of his tweet can be found at https://www.maproomblog.com/2020/07/georgias-covid-19-maps-bad-faith-or-bad-design/.

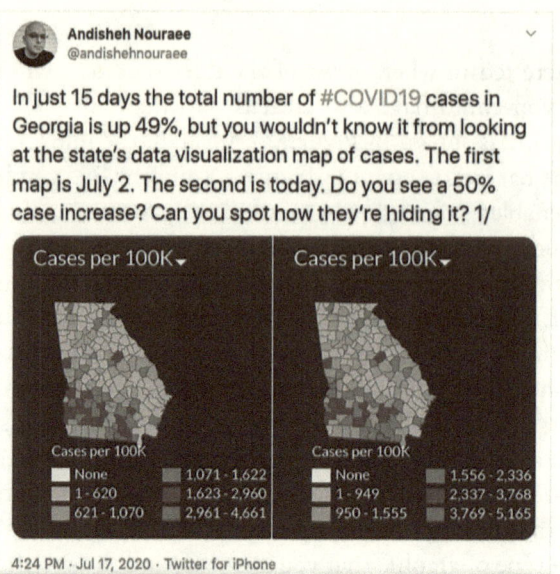

Figure 7.7: Infographics created by the Georgia Department of Health during the COVID-19 pandemic

In a local news story, a representative from the state health department responded: "These maps are not designed to show increases over time, but rather to show density by location and difference between counties."[12] The organization also acknowledged the confusion and claimed that they were in the process of redesigning the maps to make them more accessible for non-expert audiences. Yet, arguably, the intended audience for these maps—that is, the general public—**only** cares about "density by location" and not about "difference between counties." To put it bluntly, if you live in the state of Georgia and you're looking at these maps, it's for one reason and one reason only—to determine if COVID-19 cases are increasing in your area. Because you may want to use this visual information to decide what actions to take to prevent exposure to the virus such as engaging in social distancing, staying at home, getting vaccinated, and/or wearing a mask in public areas. You don't need to know whether cases are higher in your county compared to other counties in the state. The information presented in the maps in Figure 7.7 does not enable viewers to make those decisions. Thus, the maps do not effectively convey the visual information that viewers need to know to make decisions and take actions.

---

12 See https://www.wsbtv.com/news/georgia/public-health-experts-call-out-confusing-covid-19-data-maps-dph-set-make-changes/KZPBOLBG2BG2TAMI2N4CQGWESA/.

This example reinforces the importance of meeting the audience's informational needs by attending to genre selection—that is, the specific type or category of graphic—when visualizing data as well as design decisions like color choice and scale. For instance, as discussed towards the beginning of this chapter, red is alarming, and blue is calming. If the creators of these visuals wanted to convey that the pandemic was under control, blue might be an appropriate color choice. However, if they wanted to increase alarm and encourage viewers to engage in behaviors that slow the spread of the virus, using a red color scheme would accomplish this more effectively. As a working professional, you will need to make similar kinds of decisions when you visualize data.

STRATEGIES FOR USING DATA VISUALIZATIONS PERSUASIVELY

As you think through the discussion in this section, carefully consider the kinds of relationships you want to create when visualizing data that will meet your audience's informational needs. More specifically, do they need to see comparisons (e.g., bar charts and tables), trends over time (e.g., line graphics) or spatial/location information (e.g., thematic maps)? Consider how they will be using these visuals to make decisions and take actions.

To illustrate these ideas in more detail, consider another hypothetical workplace scenario. Envision that you're a visual journalist working for a large, multinational news organization in the US. It's January, and the country is experiencing a particularly severe flu season. Your organization wants to publish an article about the number of cases being reported in different parts of the country. You interview a public health expert who directs you to a map that shows flu activity by state.[13] You recreate the map and group states by the following geographic regions: Northeast, Northwest, Southeast, Southwest, Midwest. You find that the most cases are in the Southeast followed by the Midwest, Northwest, and Northeast. The Southwest has the fewest number of cases. The original map used shades of green to show areas with minimal and low activity and shades of red to show areas with high and very high activity. But you think it might be too alarming to use red for the Southeast, so you use a grey-scale color scheme (i.e., black, white, and shades of grey). You select a dark shade for that region, and you choose progressively lighter shades for the others. You use the lightest shade for the Southwest, the area with the fewest cases. You include a footnote that provides the original source for the data shown in your map. And you describe behaviors that readers can engage in to prevent getting the flu like staying home if they have flu-like symptoms and

---

13   https://www.cdc.gov/flu/weekly/usmap.htm

regularly sanitizing their hands. Your readers will use this graphic to determine if they live in an area where flu cases are high and to decide the kinds of behaviors they might engage in to reduce their chance of being infected.

Guiding Questions: When you visualize data, carefully consider the following:

- What kinds of relationships does your audience need to understand about the data—temporal (line graphs), comparative (bar graphs) and/or spatial (maps)?
- What color choices will you use in your graphics and why? What kind of emotional response do you want to elicit from your viewers (e.g., alarming, calming, something else)?
- What decision does your audience need to make and/or what action do they need to take after viewing your graphics?

Extended Example: Rural Studio—An Exploration into Citizen Architecture

This final section shows and discusses the visual choices used in two documents created by graduate students for Rural Studio, the program described at the beginning of this chapter. The first document is a brochure designed for *The Breathing Wall Mass Timber Research Project,* an innovative building system designed to "reduce environmental impact and energy use."[14] Visitors to the Rural Studio facility can pick up this brochure inside the test buildings that were built by students, which demonstrate how this technology works, or at the entrance to the program's headquarters, Natalie Butts-Ball, the communications manager for the program, explained. The second document, the Harvest Calendar, shows the growing schedule for different crops at the Rural Studio Farm. This poster was created to be displayed in the kitchen where students, staff, and faculty pick up prepared meals; it was designed to show viewers the vegetables they will be eating at certain times of the year and "directly connect them to what they are helping to grow on the farm and what is in season," Butts-Ball stated. Both documents can be found in the appendix.

To now discuss the visual communication choices used in these documents, the brochure uses all of the visual forms discussed in this chapter. The first page shows photographs of the research team members, realistic wood grain imagery behind the title of the brochure and the description

---

14 See http://ruralstudio.org/project/breathing-wall-mass-timber-research-project/

of Rural Studio, and an illustrated map[15] showing the program's location. The headshot angle of the photographs creates the least amount of distance between the subject matter and viewers, inviting viewers to learn more about the students who contributed as individuals and personalizing the brochure. The imagery in the background semiotically establishes *wood* as the major theme, while the map emphasizes the county in Alabama where the project is housed.

The second page of the brochure includes a photograph in the top left-hand corner, which denotatively conveys wood and connotatively conveys the value of sustainable building materials—also reinforced by the explanatory paragraph below this image. The main focal point is the annotated drawing in the center that illustrates how this technology works, which uses a hand-drawn illustrative style. As with the image of the human heart discussed earlier (Figure 7.3), it would be difficult to convey this information using a photograph. Rather, the abstracted drawing focuses viewers' attention on the specific process being described as beige colored arrows inside the room show the direction of airflow. The full-length human figure inside the room is climbing a ladder, but the identity of the person is not important. Rather, this visual element serves to convey the room's scale to viewers. Smaller explanatory illustrations are included around the focal point that provide further details about the process. In this imagery, orange signifies the flow of warmer air outside the structure whereas blue signifies the cooler air inside. Finally, a thematic map, which visualizes the "percent of timberland per county" for the state, is included on the bottom left-hand side with the green shading here semiotically reinforcing the environmental theme of sustainability.

Green is not a dominant color in the brochure because the focus is on wood and sustainable building. But green is dominant in the Harvest Calendar where it signifies growth and visually reinforces the theme of vegetable harvesting. More specifically, the green and grey color scheme that occupies the top portion is used to show viewers the growing schedule in two areas of the farm: the greenhouse and field. The bottom portion of the calendar then provides specific details about the crops listed above. In this section, small photographs are used to show each of the vegetables, a highly realistic choice that creates a more "vivid" representation for viewers, as Charles A. Hill would characterize it. The vegetables are shown close-up and at eye-level—as they might look if they were sitting right in front of you.

---

15 The map and original perspective section were created by graduate students in the Rural Studio program.

Final Thoughts

This chapter has sought to discuss several visual communication genres and to provide strategies that you can use to create persuasive visual forms. As with other technical communication artifacts, as you create visual communication, it's important to continually keep your audience and their informational needs and expectations in mind. Be aware that your viewers are not passive observers of visual information but rather active *users*. **They use your information to make decisions and take actions.** So, aim to create visual forms that enable them to do so.

Works Cited

Amare, Nicole and Alan Manning. "Teaching Form and Color as Emotion Triggers." *Designing Texts: Teaching Visual Communication,* edited by Eva Brumberger and Kathryn Northcut, Routledge, 2013, pp. 181-195.

Barthes, Roland. *Image, Music, Text.* Translated by Stephen Health. Hill & Wang, 1977.

Buolamwini, Joy. "When the Robot Doesn't See Dark Skin." *The New York Times.* 21 June 2018, https://www.nytimes.com/2018/06/21/opinion/facial-analysis-technology-bias.html

Butts-Ball, Natalie. Personal Communication, 3 Nov. 2020.

Hariman, Robert, and John Louis Lucaites. *No Caption Needed: Iconic Photographs, Public Culture, and Liberal Democracy.* University of Chicago Press, 2007.

Helmers, Marguerite, and Charles A. Hill. "Introduction." *Defining Visual Rhetorics*, edited by Charles A. Hill and Marguerite Helmers, 2004, pp. 1-23.

Hill, Charles A. "The psychology of rhetorical images." *Defining Visual Rhetorics*, edited by Charles A. Hill and Marguerite Helmers, 2004, pp. 25-40.

Kress, Gunther R., and Theo Van Leeuwen. *Reading Images: The Grammar of Visual Design.* Psychology Press, 1996.

McCloud, Scott. *Understanding Comics: The Invisible Art.* Kitchen Sink/Harper Perennial, 1994.

Saussure, Ferdinand de. "Nature of the Linguistic sign." *Readings for a History of Anthropological Theory,* edited by Paul A. Erickson and Liam D. Murphy, 4th ed., University of Toronto Press, 2013, pp. 87-92.

Schön, Donald A. *The Reflective Practitioner: How Professionals Think in Action.* Vol. 5126. Basic books, 1984.

Sturken, Marita, and Lisa Cartwright. *Practices of Looking: An Introduction to Visual Culture.* 3rd ed., 2017.

Teacher Resources

Overview and Teaching Strategies

Over the past few decades, visual forms of communication have increasingly been recognized as a central component of technical communication practice. More specifically, we no longer think of visuals as just 'aids' that simply illustrate or otherwise merely support the written content in a document. Rather, visual communication is its own powerful semiotic mode that works with other modes—writing, speaking, gesturing—to convey particular meanings. At the same time, visuals also construct and communicate ideas independently from these other modes. Indeed, we often encounter visuals that we have never seen before—the lead photograph in an online news story, a catchy YouTube video advertising the latest iPhone, a picture a friend just posted to social media—that we immediately understand because we are familiar with the context. And because we are unconsciously drawing from our own cultural knowledge to create that meaning.

Given the ubiquity and importance of visual communication, this essay can be taught alongside instruction in the major genres in technical and professional communication—reports, proposals, instructions, descriptions, definitions. Today, all of these documents are likely to use visuals to convey concepts and information. This chapter explains to students how several commonly-used visual forms work and provide strategies for taking a rhetorically informed approach in their composing practices. The main take-away then for students reading this chapter is to attend to creating visuals with the same level of effort and thought that they give to their writing. More specifically, students should assess the rhetorical situation. That is, they should carefully consider the audience for the visual, how the audience will use it to make decisions and take actions, and the interpretive context in which it will be viewed. Engaging in such rhetorical decision-making will help students determine how they can create visuals that will be effective and persuasive in meeting the information needs of their readers.

Discussion Questions and Activities

The following questions and activities prompt students to engage with the material in this chapter and can be done individually or in groups and then discussed together as a class.

1. You have likely heard the well-known expression that this chapter draws its title from: "A picture is worth a thousand words." Consider the differences between how we interpret *pictures* (i.e.,

imagery and other visuals) and how we interpret *words*. What can images convey that words cannot and vice versa? Are there areas of overlap? What might these be? Consider the assignment you are currently working on for this class and determine what information you need to present in writing and what information would most effectively be presented visually. Why? Be prepared to share your reasoning.

2. Find an advertisement (digital or printed) that has few if any words and *read* the ad using the information in the semiotics and/or photography sections of this chapter. What visual signs do you see, and what do they mean? How do you know? If the ad uses pictures, why do you think the photographer chose the particular subject matter selected? How are you interpreting the camera angles used and the composition of the photograph? How do these choices add to the meaning of the ad?

3. Envision you are a manager at an investment firm. Your supervisor has asked you to prepare a report that analyzes the financial portfolios of your firm's four most profitable clients. Visualize the data provided below that gives percentages for the portfolio for each company. Explain why you chose to visualize the data using a particular format and compare your visualization with those created by your peers.
   - **Company A:** mutual funds, 18.5; annuities, 32.8; real estate, 3.1; stocks, 1.2; bonds, 44.4.
   - **Company B:** mutual funds, 20.3; annuities, 26.9; real estate, 3.8; stocks, 4.0; bonds, 45.0.
   - **Company C:** mutual funds, 23.5; annuities, 23.8; real estate, 7.3; stocks, 4.1; bonds, 41.3.
   - **Company D:** mutual funds, 20.3; annuities, 25.2; real estate, 9.6; stocks, 6.3; bonds, 38.6.

4. Find a short set of instructions (between 6-8 steps) that uses visuals that are difficult to understand. Assess the rhetorical situation and explain why the visuals are difficult to understand. Then redesign the visuals to address the issues you identified.

5. Envision that you want to buy a new smartphone, laptop, or tablet. Make a list of 3-4 criteria that are important to you in making this decision: cost, features, design, battery life, etc. Research your criteria and then create a visual that allows you to compare what you found. What product do you think you will choose based on the information in your graphic?

APPENDIX

TECHNICAL WRITING SPACES

RESEARCH TEAM

Jacob Elbrecht
Lynchburg, Virginia

Jake believes architecture's purpose is to create space, and all other elements of architectural design should serve this purpose. To this end, he sees good design as timeless, while styles and trends lose their value over time. He can be found in Rural Studio computer lab all hours of the day and night handling our IT problems.

Katherine (Fergie) Ferguson
Nashville, Tennessee

Katherine, more commonly known as Fergie, has a down-to-earth nature and positive outlook that are mirrored in her architectural endeavors, in which she strives to create beautiful and dignified spaces for all people. This architectural philosophy stems from her "live and let live" attitude and love of the natural beauty of the world.

Anna Halepaska
Dallas, Texas

Anna believes architecture possesses a type of beauty that can have a positive impact on the world beyond the visual. She brings a bubbly and unique personality to the team with her affinity for teal and love of plants. For her, Rural Studio is a chance to use her passion to continue to construct hopeful contexts for Hale County.

Preston Rains
Birmingham, Alabama

Preston views everything as a design challenge, from architectural work to culinary experiments. For him, Rural Studio is an exploration of the tactile value of design, and he hopes to augment his design and construction repertoire while growing a luscious beard.

LEARN MORE

What kind of building does this work for? Can this be installed in a residential home?
The system isn't developed enough yet to say for sure, but it seems like it would pair well with a building typology with a relative occupant load such as an office building.

Does it have to be timber?
No. The Breathing Wall system can theoretically work with any solid material. Timber turned out to be the best choice for our research.

Does this work in the summer?
This research is only addressing a heating scenario, but the system could be flipped to cool the interior of the building instead.

How big are the holes and how are they oriented?
The size of the holes changes depending on the design of the specific wall. They can range from minuscule to up to half an inch. The holes are oriented perpendicular, running them interior to exterior.

How are the walls held together?
These Tall Buildings use a timbered nail construction method, where each piece of wood is attached to the top of next using nails hammered down with a ¼" Strill nail. No glues are used in building of the clusters that the timbers themselves. It is easily disposable at end of life cycle from each face.

Can you feel the airflow?
No. All Breathing Wall research at Greene Commons available to anyone who is passionate about carrying this technology forward.

Where does the timber come from?
Our timber comes from a local lumber mill, and is harvested within a 150-mile radius from our location.

AUBURN UNIVERSITY

RURAL STUDIO
AUBURN UNIVERSITY
Newbern, Alabama

BREATHING WALL
MASS TIMBER RESEARCH

ABOUT RURAL STUDIO

Rural Studio is an off-campus design-build program, part of the School of Architecture, Planning and Landscape Architecture of Auburn University. The program, established in 1993 by D.K. Ruth and Samuel Mockbee, gives architecture students a hands-on educational experience while assisting under-resourced communities of West Alabama's Black Belt.

The Rural Studio philosophy suggests that everyone, both rich and poor, deserves the benefit of good design. In its initial years, the Studio became known for establishing an ethos of recycling, reusing, and remaking. In 2001, after the passing of Samuel Mockbee, Andrew Freear succeeded him as director. To fulfill this ethic, the Studio has expanded the breadth and scope of its work to address community need. Projects have become multi-year, multi-phase efforts across five counties. The students work within the community to define solutions, fundraise, design and ultimately, build remarkable projects.

The Studio continually questions what should be built, rather than what can be built. To date, Rural Studio has built more than 200 projects and educated more than 1,000 "Citizen Architects."

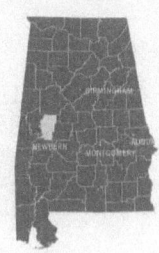

BREATHING WALL MASS TIMBER PROJECT

Breathing Wall Mass Timber research investigates designing the building envelope as a heat exchanger in mass timber construction. The Breathing Wall maximizes the thermal, structural and ecological properties of wood to design an envelope as a heat exchanger so that incoming fresh air can be tempered with low-grade heat.

The final output of this work is two test buildings which represent the culmination of knowledge gained from a series of smaller experiments concerning the transfer of the Breathing Wall concept from a laboratory scale to building scale.

WHAT IS A BREATHING WALL?

By introducing a mathematically optimized grid of holes into a massive, monomaterial envelope, the Breathing Wall technology allows that wall to act as a heat exchanger, thus transferring the heat stored within the wall to the air moving through the grid of holes. Unlike traditional insulative walls which leak heat at a rate related to their insulative value, the Breathing Wall actively recovers heat, allowing it to act much more insulative than its constituent elements would on their own.

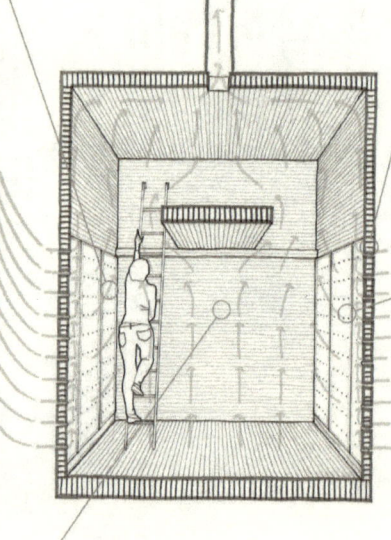

SUSTAINABLE CONSTRUCTION

Due to timber's renewability and carbon sequestering capabilities, it is the ideal material for a single material building. If harvested responsibly, a mass timber building can remove huge amounts of carbon from the atmosphere and store it for many years.

WHY MASS TIMBER?

The Breathing Wall system requires a thick wall in order to function effectively. Mass timber not only provides that but also creates a more thermally comfortable and environmentally friendly space than many other materials.

HEAT EXCHANGER

Most walls act as an insulator but the Breathing Wall acts as a heat exchanger. It takes heat from the exterior walls of the building that would otherwise be lost and passes it back into the air coming into the building.

THERMALLY ACTIVE SURFACE

A Thermally Active Surface (TAS) is a conductive/convective heating element that is placed directly on the interior surface of the Breathing Wall and provides the heat required to drive both the heat exchange and buoyancy ventilation of the Breathing Wall.

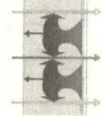

BUOYANCY VENTILATION

Buoyancy ventilation refers to airflow generated by warm air's tendency to rise and is required for the heat exchange effect to function. The rising warm air draws cooler air through the holes of the Breathing Wall where it interacts with the heat stored in the wall, thus driving the heat exchange effect.

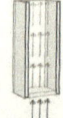

CONNECT:

334.624.4483 | rstudio@auburn.edu

Auburn University Rural Studio
8448 Alabama Highway 61
Newbern, Alabama 36765

www.ruralstudio.org | www.ruralstudioblogs.org

@rural.studio
@ruralstudio
@RuralStudio

Auburn University is an equal opportunity educational institution/employer.

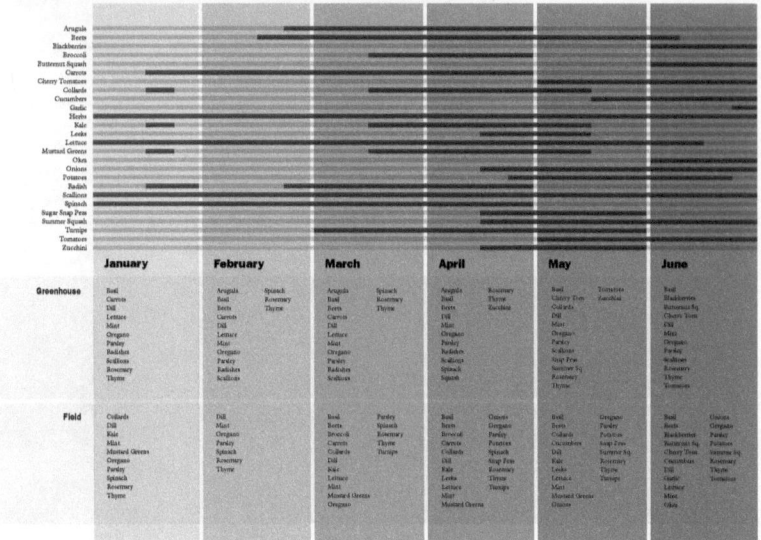

8 Technical, Scientific, and Business Presentations: Strategies for Success

Darina M. Slattery

Throughout your time in college, you are likely to be asked to develop presentations, either on your own or with project team members.[1] While many students find giving presentations stressful, the more practice you can get the better, because you are likely to have to give one or more presentations in your professional life. For example, if you end up working as a scientific or technical writer, you might have to give a presentation to the project manager about the status of a documentation project you are working on (perhaps some of your colleagues are not submitting their drafts on time, or perhaps you have identified some inconsistencies in writing style that need to be addressed). Or, if you end up working as a training needs analyst in a large technology corporation, you might have to give a presentation to management about employees' training needs, while also outlining some solutions that will help meet those training needs.

As a technical or scientific writer, you will need to be able to prepare different kinds of documents to meet the needs of audiences who have varying needs and goals. Presentations are often extensions of other technical writing 'products,' such as technical reports, product specifications, and user manuals, so you will need to think carefully about the purpose of your presentation and adapt the content, and your approach, accordingly. For example, a technical report about a new vaccine that is written for people in medical fields would need to provide detailed information about the vaccine (the ingredients, possible side effects, and the results of the clinical trials), whereas an oral presentation about that technical report might only summarize the *key* findings from the clinical trials, if the goal was to encourage the general public to get vaccinated. In other words, you need to

---

1 This work is licensed under the Creative Commons Attribution-NonCommercial-NoDerivatives 4.0 International License (CC BY-NC-ND 4.0) and is subject to the Writing Spaces Terms of Use. To view a copy of this license, visit http://creativecommons.org/licenses/by-nc-nd/4.0/, email info@creativecommons.org, or send a letter to Creative Commons, PO Box 1866, Mountain View, CA 94042, USA. To view the Writing Spaces Terms of Use, visit http://writingspaces.org/terms-of-use.

provide different levels of detail, and present your content differently, in an oral presentation that is based on a technical report.

This chapter will outline the key phases of presentation development and delivery, and how best to go about them, with reference to theory and practice from the disciplines of technical communication and information design. While the purpose of this chapter is to prepare you for workplace presentations, some of the examples will refer to classroom presentations too, as that is where you will initially develop your presentation skills.

The key phases discussed in this chapter will include:

- Identifying the audience and the purpose of the presentation
- Choosing presentation tools
- Developing the presentation
- Delivering the presentation

These phases are key to developing effective presentations, because they allow you to address the needs of the audience (who they are and what they need to hear/learn), choose appropriate tools to convey the message of the presentation, apply best principles to the design and development of that presentation, and deliver the presentation in an effective way. As a presentation will often accompany other forms of documentation, the other skills you learn throughout this handbook will also help you to fine tune your presentation skills.

Identifying the Audience and the Purpose of the Presentation

In this section, we will consider the following ideas:

- Audience demographics
- Audience interests
- Audience expectations
- The purpose and scope of the presentation

This focus is key, for before you can begin to develop a presentation, you need to find out as much as possible about your target audience. Are you presenting to peers who have similar interests and abilities to you, to your professors, or to workplace colleagues?

If you are giving an update to teammates about the status of a group project, the presentation will likely be informal, as you will just be addressing the key points that your teammates need to know. If your presentation is for a professor and is credit-bearing, it will need to be more formal, partly because your professor will be evaluating the quality of your presentation and delivery style. Your professor will likely specify topics that you need to address and may even issue a grading rubric to help you plan your presentation. To achieve a high grade for a college presentation, you should ensure you adhere to all the requirements for the presentation and that you address each of the specified topics.

In a workplace setting, where you may be presenting to peers or to senior management, you need to remember that your colleagues might not have the in-depth knowledge that you have about what you do on a daily basis. Consequently, you will need to think carefully about the language you use, the tone you adopt, and the content you cover. In a presentation to management, the audience might only be interested in specific topics. For example, in a software company, management might only want to know whether or not you will complete the software documentation on time and might not be interested in *how* you went about designing that documentation.

Throughout your college and professional life, you might hear people say 'know your audience'—this is because we sometimes forget about our audience and give the presentation *we* would like to hear, rather than the one the audience would like to hear! We will talk more about content, and how best to present it, in a later section of this chapter.

CHOOSING PRESENTATION TOOLS

In this section, we will consider the following factors:

- The available software you can use for creating or giving presentations
- Your technical proficiency (skills) with such software
- Whether or not you are using a free trial version of the software
- Output file formats available from different software

Such factors are important, for there are many commercial applications on the market, as well as free-to-download applications that you can use to develop presentations. Most tools offer similar features, including the ability to insert text, images, and other media, and to add special effects such as transitions (e.g., slides that fade in/out) and animations.

Two commonly used tools include Microsoft PowerPoint (commercial) and Google Slides (free-to-use), but there are many others. You might already have access to one or both of these tools on-campus or on your personal device.² In addition, most workplace environments provide employees with access to a suite of tools that will help them undertake their day-to-day duties, so it is likely that a presentation tool will be in this suite.

While there are many elaborate tools out there, you should consider the tools you have used previously (if any) and use one of those, unless you have time to learn new software. You will find lots of helpful 'how to' videos for commonly used tools on YouTube and other websites, so it's best to use one of those if you have never used presentation software before. For example, if you type 'How to insert a YouTube video into Google Slides' or 'How to add effects in PowerPoint' into Google, you will get a large number of results. Tools like Prezi can offer appealing effects, but the format might not work as well in a corporate environment, and besides, you (the presenter) would need to be comfortable using Prezi, so that is another consideration.

If you need to use software that you don't currently have a license for, you may be able to download a free trial version. However, it's important to note that these tend to have time limits (typically 30-60 days) and/or feature limitations (e.g., you might not have access to the full range of features or there might be limits on the number of presentations you can save). But quite often the free trials will suffice for college presentations, which typically only span a few weeks. For workplace presentations, you should ideally have access to the full version, with no feature limitations, so you will need to check this out with your I.T. department.

> **Useful tip:** Don't install a trial version until you are ready to start working on the presentation and make sure the trial license will remain valid until your deadline.

You need to find out if your target audience (this could be your professor or an employer) has requested specific file formats. Typically, Microsoft PowerPoint presentations have a ppt or pptx filename extension, but sometimes you may be asked for a pdf file, which can be viewed in a Web browser. Fortunately, PowerPoint enables you to export ppt files as pdf files, so that won't be an issue. Google Slides also enables you to save presentations in ppt or pptx formats. If you are giving a live presentation, you should also ensure that the software you have used will work in the presentation venue—I've attended a number of presentations over the years where the presenter did not check this beforehand and was unable to show their presentation slides (also known as a slide deck). We will talk more about technology considerations later in this chapter.

---

2 You can read more about Microsoft PowerPoint at https://www.microsoft.com/en-us/microsoft-365/powerpoint. You can access Google Docs at https://docs.google.com.

> **Useful tips:**
> - Check out the room facilities a few days before your live presentation to make sure you will have access to the software you are using. Ideally, do a trial run beforehand, to ensure your presentation displays as intended. You don't want any surprises on the day!
> - If you cannot access the venue before your live presentation, save your file in a variety of common formats (e.g., ppt and pdf) to give yourself a greater chance of your presentation working as intended.
> - Bring a copy of your presentation with you on a USB (pen) drive but also email yourself a copy in case the USB drive fails to work. Being prepared will also boost your confidence.

Developing the Presentation

In this section, we will consider the following questions:

- How should you structure technical, scientific, and business presentations?
- What are the considerations for writing style and tone?
- How should the information be designed, in terms of text formatting, use of color, and incorporation of tables, charts, and diagrams?
- What are the advantages and disadvantages of using presentation templates?

Let's look at each of these topics in a bit more detail.

Structuring the Presentation

A presentation typically comprises three sections: an opening, a body, and a closing. In the opening, you should first gain the attention of your audience by providing some context for the presentation you are giving. For example, if you are giving a presentation to your professor about your team's brochure redesign assignment, you might present a screenshot of the finished brochure on the opening slide. In the opening section, you should also outline what the presentation will cover—think of this outline slide as a 'table of contents' for what will follow and present it in bullet points, rather than in paragraph format. The goal of the opening is to capture the attention of the audience and let them know what you are going to address in your presentation.

In the body of your presentation, you present the main message. Continuing with the college brochure assignment as an example, you might dedicate separate slides to discussing why your team picked that brochure, why they redesigned it that way, the challenges the team encountered, and what the team would do differently in future. In a corporate presentation to management, you might present an update on how the documentation project is progressing, which deadlines have been met, which challenges still remain, and so on. Aim to have a single message on each slide and state it clearly. While bullet points are good for lists, they are not suitable for every message. We will talk more about information design in the next section.

In the closing section, you should wrap up your presentation by summarizing the key points of the presentation. Students sometimes struggle with summaries, as they're not sure which points to repeat and which ones to leave out. Imagine that your audience will only look at the summary—what are the key points (findings) they must take away? What lessons will they learn? What are the next points of action? You also need to make it clear to your audience that you have finished the presentation, to avoid any awkward moments of silence at the end. One way to do this is by closing with the statement 'Thank you for your attention. Have you any questions?' We will talk more about delivery and how to wrap up a live presentation, in a later section.

Writing and Formatting the Presentation

Once you have picked your presentation tool, start laying out your slide deck. Many tools offer templates, which are pre-defined designs that you can customize to your own use. Templates are really useful if you are caught for time, do not have good design skills, or just want to get inspiration. Usually, the templates and design ideas are informed by good practice, so you can be reasonably confident that the type style, sizes, and colors work well together.

> **Useful tip:** Presentation software templates are a great starting point if you are looking for inspiration about colors and typography that work well together. Even if you're not creating a presentation, you can look at the color schemes and layouts that have been built into the templates to get ideas for your other documentation projects.

While templates are always an option, you might prefer to design your own layout; if so, you should keep the following guidelines in mind:

First, carefully consider the type style and size you use on your slides. Sans-serif types (like Verdana, Century Gothic, and Arial) tend to work better for on-screen presentations as they don't have curves at the end of each letter and are therefore more uniform (unlike serif types like Times New Roman and Century Schoolbook) (Schriver 252–256). The subject matter can also influence the type you choose—for example, a sans-serif type like Comic Sans would not be appropriate for a workplace presentation. The type style will also dictate the size of the type—some types look larger than others, so you may need to experiment a little.

Next, avoid using italicized type (except for very small portions of text) as it is difficult to read and avoid using underlining completely (except for hyperlinks) as it distorts the type (Schriver 265–266). Avoid using all uppercase letters as it reduces reading speed and can come across like you are shouting. Also, only use a numbered list when the text order is important (e.g., for a set of procedural steps) and a bulleted list for all other lists.

Finally, limit the amount of text you place on each slide. You may have heard the phrases 'less is more' and 'keep it simple'—well those principles equally apply to presentations. If possible, only include keywords or themes, and talk around them as you address the audience. Try to avoid paragraphs of dense text, as attendees will find them difficult to follow (see Figure 8.1), but if you must use paragraphs, chunk ideas into separate paragraphs. Left-align paragraphs of text.

As discussed earlier, the tone you adopt and the language you use should be informed by the audience you are addressing. If the audience comprises people who might not be experts in your discipline or field, or who may be non-native English speakers, you should avoid using jargon, colloquialisms, clichés, and acronyms.

> **Useful tip:** Plain English is a style of writing that ensures the content is more readily understood by everyone. You can read about the Federal Plain English Guidelines at https://plainlanguage.gov/guidelines/.

In Figure 8.2, note how a table has been used to present essentially the same information as Figure 8.1, but in a more user-friendly way. It is better to use a table to present multiple numerical values as you can present the information in a relatively small space and the audience will find it easier to compare the values (Rubens 336-337). A table will also help you during your presentation as you won't have to remember the data and can elaborate on what the data means, in your own words.

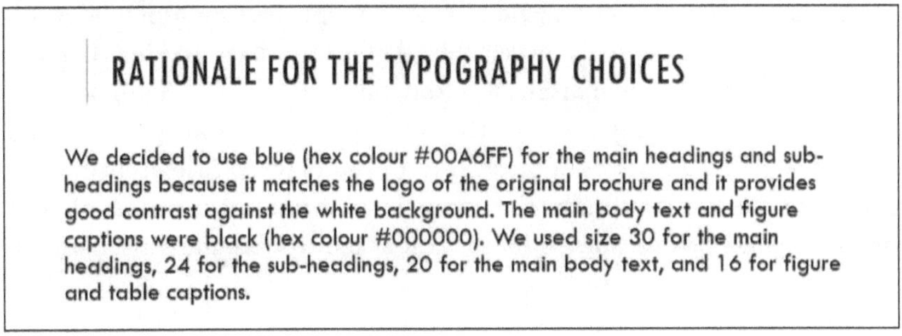

RATIONALE FOR THE TYPOGRAPHY CHOICES

We decided to use blue (hex colour #00A6FF) for the main headings and sub-headings because it matches the logo of the original brochure and it provides good contrast against the white background. The main body text and figure captions were black (hex colour #000000). We used size 30 for the main headings, 24 for the sub-headings, 20 for the main body text, and 16 for figure and table captions.

Figure 8.1: Example of a Poorly Designed Slide

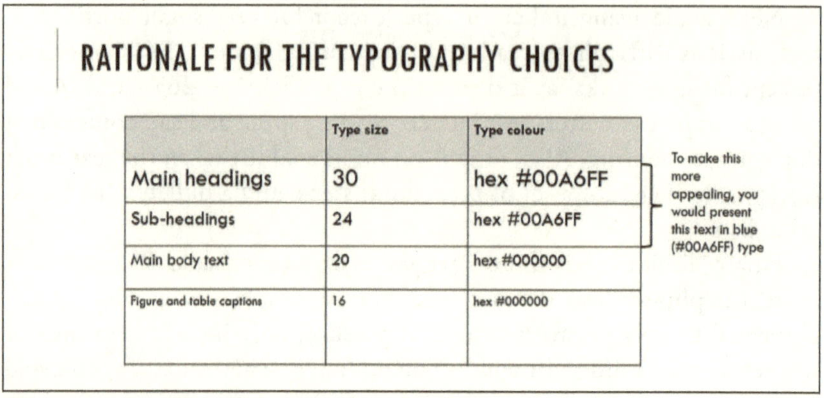

RATIONALE FOR THE TYPOGRAPHY CHOICES

	Type size	Type colour	
Main headings	30	hex #00A6FF	To make this more appealing, you would present this text in blue (#00A6FF) type
Sub-headings	24	hex #00A6FF	
Main body text	20	hex #000000	
Figure and table captions	16	hex #000000	

Figure 8.2: Example of an Improved Slide Design

You can often use figures, diagrams, and charts to explain complex concepts and to appeal to attendees who prefer visuals to text. If you are presenting the results of a survey, for example, you might use a pie chart or bar graph to convey the data, rather than present the figures on their own. If you are presenting a workflow or process, a flowchart can work well.

Every visual should be simple, similar in size and placement, and high quality. Callouts or labels should be legible. Include informative captions for every table, figure, graph, chart, and graphic you incorporate into your slide deck. Text and visuals should complement one another (Schriver 441).

When choosing graphics, try to avoid using clipart (particularly in workplace presentations) as they tend to look unprofessional; instead, look for high quality images. Two sources of possible images include Pexels and Creative Commons.[3] You might occasionally want to incorporate audio or video clips into your presentation. If you do, make sure that they are relatively short clips and that they will add value to your presentation.

Use blank space as a design feature. Fortunately, it doesn't cost any more to add another slide (unlike printing additional pages in a hard copy of a document!) so don't be afraid to spread a topic over multiple slides. It's easier on the eye if there is a healthy blend of text, visuals, and blank space, so be generous with the margins too (top, bottom, left, and right).

You should carefully consider the colors you use in your presentation. If you are giving a classroom presentation, you may have flexibility in the colors you use, or you might find some inspiration in the theme of your presentation. For example, a presentation about a brochure assignment

---

3 Pexels can be found at https://www.pexels.com. Creative Commons can be found at https://search.creativecommons.org.

might incorporate similar color schemes to those used in the brochure. In a workplace setting, the corporation might already have pre-determined templates that you must use, which will usually be aligned with other corporate materials. But in the absence of corporate templates, you should use professional looking templates that use *appropriate* type styles and colors.

> **Useful tip:** If a topic needs to span more than one slide, you can use a convention like this for the slide headings, where '1 of 2' means the first slide of 2 on that topic:
> - Rationale for the typography choices (1 of 2)
> - Rationale for the typography choices (2 of 2)

Regardless of templates and corporate guidelines, you should always ensure that your color choices provide sufficient contrast. Black text on a white background provides maximum contrast, but it can look very stark, so you might like to consider dark shades of grey or navy type on an off-white, cream, or pale-yellow background. You may need to experiment with a number of color combinations to get it 'just right'. You should also bear in mind that some colors have cultural connotations that will deem them inappropriate.

You should also try to make your presentations accessible to as many attendees as possible. Color choices can also cause accessibility issues. According to the Colour Blind Awareness organization, about 300 million people worldwide are color blind. Some colors (including red and green) cannot be perceived by people who have color blindness.

For attendees who are blind, you can provide a text transcript of your content so their screen reading devices can read the text to them. In addition, by providing alternative (alt) text for images, figures, or tables, your attendees will hear what you are trying to convey in the image, figure, or table, as their screen readers will read the alt text to them.

Attendees with dyslexia will benefit greatly from appropriate font choices—sans serif typefaces are preferable to serif typefaces, as they are plain, evenly spaced, and more readable, especially if they are size 12 or larger. Other formatting features that will help readers with dyslexia include left-justified text (the text should be jagged on the right), short paragraphs, 1.5 line spacing, and more visuals to ease the reading load.

Finally, you can help attendees who are hard of hearing by wearing a microphone and speaking clearly and slowly. Attendees who are deaf will benefit from receiving copies of your presentation notes in advance, so they can follow the presentation in text form.

> **Useful tips:**
> - This article outlines color and cultural design considerations: https://www.webdesignerdepot.com/2012/06/color-and-cultural-design-considerations/.
> - You can read about color blindness on this website: https://www.colour-blindawareness.org/colour-blindness/.
> - You can test your own color vision on this website: https://enchroma.com/pages/color-blind-test.
> - Many presentation tools can check for common accessibility issues such as missing alt text and poor contrast between foreground and background. Look for 'accessibility' features.
> - This website, by the World Wide Web Consortium, offers useful tips for writing, designing, and developing accessible content: https://www.w3.org/WAI/design-develop/.

Delivering the Presentation

In this section, we will consider the following questions:

- What are the most important technology and room facility considerations when delivering a synchronous, in-person presentation?
- Why is it important to do rehearsals before a live presentation?
- What are the different styles of delivery?
- How can presenters plan for interaction?
- How should you use supplementary resources and follow-up with attendees?
- What are the additional considerations when delivering a synchronous, virtual presentation?

These factors are key to giving effective presentations that address many of the items noted in the previous sections of this chapter.

Technology and Room Facility Considerations

As mentioned earlier in the section on choosing tools, you need to make sure that your presentation will run in the presentation room, as not every PC has the same setup as yours. To give you time to sort out any unexpected issues, you should test-run your presentation a few days before the live presentation. You should also ensure that you know how to operate the overhead projector, as some venues use remote controls to operate overhead displays, and they can take a little getting used to.

If you cannot get access to the presentation room a few days beforehand, go to the room early on the day of your presentation to set everything up and make sure you bring a few versions of your presentation with you, as mentioned earlier (e.g., ppt and pdf).

Ideally, you should also do a sound check before the live event by asking a colleague to stand at the back of the room as you speak—they will be able to tell you if the volume is sufficient or if you need to borrow a microphone. If you decide to use a microphone, make sure you have tested it beforehand and are familiar with it.

If you are comfortable using a pointer device, you might want to consider using one. These devices usually come in two parts—one part plugs into the USB drive and you hold the other in your hand. You can use the pointer to highlight areas on the screen (they usually have a red LED light) and also to progress through your slides—this latter feature is particularly useful if you need to move around the room during your presentation, as you won't have to rush back to the PC to advance the slide deck every time. We will talk more about styles of delivery later in this chapter.

REHEARSING THE PRESENTATION

I have been giving presentations at conferences for twenty years, but I still feel nervous beforehand. I even feel nervous before giving a lecture, but I usually know *that* audience, so the nerves tend to dissipate sooner! I find the best way to deal with stage fright is to spend time developing the presentation and practicing it multiple times.

It is important to time your presentation also, to make sure you have enough to say but also that you can cover all the content in the time allocated. On a few occasions, I've been surprised how quickly the time goes by and how my seemingly short presentation is actually quite long once I start talking about the content on each slide. After a few rehearsals, you will have a much better idea of what you can, and what you cannot, cover and you will feel more confident that you will be able to give an effective presentation. If you don't rehearse, you are likely to rush through the presentation and leave out important points.

You should also factor in time for questions at the end. Sometimes, the audience will not have any questions, but it's important to allow some time, just in case. If you are giving a presentation in college, your instructor will likely tell you how long the presentation must be and how much time you must allocate to Q&A. In a workplace setting, you should always allow time for questions, as it's likely that you are trying to convince your audience that your proposal or approach is a good one.

STYLES OF DELIVERY

Have you ever attended a presentation where the presenter read their slides from start to finish? Quite often, presenters know they shouldn't do this, but they are either too nervous to address the audience directly or they cannot remember what is on each of the slides (sometimes it is a mixture of both!). If you are going to read the slides out word-for-word, you might as well just distribute the slides to the audience by email, rather than invite them to the presentation venue.

There are many different presentation styles. I like to be very well prepared and use my slide deck to guide the flow of my presentation. Other presenters like to be less structured and adapt their presentation to suit the audience on the day (they might, for example, tell anecdotes to illustrate concepts or ideas). Whichever approach you take should depend on your personal communication style and confidence with the subject matter. I would advise against using an approach that doesn't suit you personally!

If you have practiced extensively, you might feel comfortable delivering the presentation from memory, just referring back to the slides as you progress through the presentation. However, if you feel the need to have notes in your hand, that is perfectly acceptable, provided you don't read them from start to finish. If you find yourself having to consult them regularly, take time to make eye contact with the audience as often as possible.

> **Useful tips:**
> - At the start of the presentation, ask the attendees if they can all hear you properly. If some people are having difficulty hearing you but you don't have a microphone, you could invite them to move further up the room if seats are still available; if not, you could present from a mid-point in the room to enable more people to hear you.
> - Don't forget to breathe during your presentation. It will make you feel more relaxed and gives you time to gather your thoughts.
> - Feel free to take sips of water whenever you need to quench your thirst.
> - If you plan to consult your notes during the presentation, make sure the pages are numbered and correctly sequenced, so you can quickly find the page you need when you need it.
> - This website presents the eight types of presentation styles: https://blog.hubspot.com/sales/types-of-presentation-styles.

Some presenters like to 'own the room'; in other words, they like to walk around and interact closely with attendees. Other presenters (myself included) tend to use the podium as their base and move from time-to-time around the top of the room, not venturing too far from the base. Either approach is acceptable once you make sufficient eye contact with attendees. For example, if you are giving attendees an overview of a software application and then want them to try the software themselves, you might walk around the room and offer one-to-one advice. It is perfectly acceptable to adopt different styles during your presentation also.

Planning for Interaction

I mentioned earlier that you should always allow for Q&A at the end of your presentation. Some presenters also like to give attendees the option of asking questions *during* the presentation—if this works for you, I suggest you let the attendees know at the start that you are happy to answer questions at any point during your presentation. If you would prefer attendees to wait until the end, you should also let them know. If you are not very confident giving the presentation, it might suit you better to leave all questions until the end, to avoid distractions. When someone asks a question, it is good practice to repeat the question, as some members of the audience might not have heard it; this strategy also helps you confirm that you have heard it correctly, before you respond.

If you need the audience to undertake some tasks or respond to your questions during the presentation, you need to make sure they have everything they need. If you are planning on distributing handouts, make sure to ask the event organizers how many people are likely to attend (and bring 10-15 spare handouts just in case!) If attendees need to undertake a task at a computer, make sure they all have logins and passwords for that venue and that the software you need is installed on those PCs.

You should also factor in the time required for interactive activities. What might take you five minutes to do might take the audience much longer, so allow extra time for those tasks.

One of the biggest concerns that most presenters face is dealing with a difficult question. Rest assured that every presenter has been faced with difficult questions at one point or another. Over the years, I've learned two things from those experiences: 1) it's ok not to know the answer to every question, and 2) you should admit that you don't know the answer but that you will find out. I find that attendees are usually satisfied once you say you will get back to them afterwards with the answer (but don't forget to do this!). Sometimes, you will remember the answer when your presentation is over, when you are more relaxed, but that's ok too.

Using Supplementary Resources and Following Up

Quite often, presenters like to distribute handouts during or after their presentations. You might, for example, like to distribute a hard copy of your slide deck, so participants can write notes on them during your presentation, or consult them later. As mentioned in the previous section, make sure you ask the event organizer how many people are likely to attend and bring spare copies. If you cannot print out a copy for attendees beforehand, you can offer to circulate them afterwards via email. Circulating them afterwards also enables you to follow up on any outstanding questions that could not be addressed due to time constraints.

> **Useful tips:**
> - Most presentation tools enable you to print your slide deck in different formats e.g., one slide per page or two to six slides per page. While it will save paper if you print a large number of slides per page, you need to ensure that attendees can still read the content on the slides. I usually print two to four slides per page to ensure they are sufficiently legible.
> - If someone else is organizing the presentation, ask them to take an attendee list so you can send the supplementary resources afterwards. If not, have a sheet of paper ready and ask attendees to add their details (names and email addresses) and pass it around, during your presentation.

Synchronous Virtual Presentations

Sometimes, due to travel restrictions or simple convenience, you will be asked to give a virtual presentation (also known as a webinar or videoconference), where you will not meet the attendees physically. While many of the guidelines that apply to face-to-face presentations also apply, there are additional considerations you need to bear in mind for a virtual presentation.

Firstly, you need to find out which software application you are expected to use—examples include Zoom, Google Meet, WebEx, Skype, and Microsoft Teams, but there are many more. If you are lucky, the event organizer will send out a 'meeting invite' by email, which will be addressed to you and all the attendees. On the day of the event, presenters and attendees will click on a link to join the virtual presentation. However, it is important that you try out the software beforehand to ensure it works as intended, as sometimes it is necessary to download an app or create a user account before you can join the presentation. On a related note, you will also need to let attendees know if they need to download anything in advance.

Secondly, like with face-to-face presentations, you should do a trial run with the software, to ensure your voice can be heard, your slide deck can be uploaded, and that your video works (if you are planning on using a webcam). Most webinar tools offer multiple ways of presenting content (e.g., slides only, slides with the presenter in thumbnail view, or presenter view only) so you will need to familiarize yourself with the various options and settle on one that works for your presentation style.

Thirdly, in terms of interaction, you need to think carefully about how your attendees will ask questions and communicate this to them. Do you want to mute all attendees by default and only unmute an individual if they have a question? Alternatively, will attendees post their questions in the chat area (most webinar tools have a text only area where attendees can converse during the event), will they put up their hand physically (on camera), or will they use the 'hand' icon (most webinar tools have a feature that enables attendees to place a small hand icon beside their name, if they want to ask a question). In my experience, monitoring the chat area and hand icons is really challenging during a live presentation, so I don't recommend it—instead, I recommend you ask a colleague to monitor who wants to ask a question and the order in which they asked them, so you can address the questions in the correct order at the end. However, if no one is available to help you, you can ask attendees to refrain from posting or asking questions until the end and then work through them one by one. And don't forget that you may need to unmute attendees if they cannot do it at their end!

Fourthly, you should also decide in advance if you want attendees to turn on their cameras or leave them off. If there are bandwidth issues, you might need to ask attendees to leave their cameras off until question time; however, the downside to this is that you may feel like there is no one in the audience, as you will not see their faces—you will likely only see thumbnail photos (if you are fortunate) or the initials of each attendee (if you are less fortunate). This kind of presentation environment takes some getting used to but rest assured that you will become more used to it in time.

Finally, common issues that presenters run into during virtual presentations include faulty microphones and webcams, so make sure yours are working and try to have backup equipment if possible. You also need to familiarize yourself with the mute/unmute feature, as sometimes presenters start speaking but the audience cannot hear them (note that you might have a mute button on your microphone as well as in the presentation tool itself). Presenters also sometimes have issues sharing their slide decks with attendees, so make sure you are comfortable sharing and un-sharing files.

Useful tips:
- Do a few trial runs of the presentation software with a colleague, to familiarize yourself with how the tool works. Delivering a presentation online can be daunting, and it can feel as though there is no one present if the cameras are off, so you need to be comfortable.
- Make sure you are familiar with the mute/unmute features.
- At the start of the presentation, ask the attendees if they can hear and see you. Also ask them if they can see your slide deck once you share it with the audience. If their cameras are off, they might answer via the chat area, so keep an eye on chat too.
- Have your slide deck open in the background and close any unnecessary windows and folders just in case you accidentally share a screen you don't intend on sharing. This happens to nearly everyone at some stage, and it can be embarrassing if you are not prepared for it.
- Put your cell phone on mute and place a 'do not disturb' sign on your door, so no one disturbs you during your presentation.
- Communicate clearly with attendees about the strategy for asking questions. If you'd prefer them to post questions in chat only and/or to hold all questions until the end, then state that clearly at the beginning of your presentation.
- If you cannot mute all attendees before they join, then ask them to mute at the start of the presentation. It can be very distracting when attendees unmute their microphones (sometimes accidentally) while you are presenting, so you need to be ready to deal with it.

Conclusion

In this chapter, we examined the key phases of presentation delivery and outlined how you might go about them in practical ways. We outlined how technical and scientific presentations are often extensions of other types of technical writing products but often have a different audience and goal. While presentations provide a different level of detail than the documents from which they originate, they must still mirror the original document. Furthermore, many of the design principles and guidelines that apply to other types of technical documents also apply to presentations.

The appendix presents a checklist that addresses the main areas you might like to consider when designing and developing your own presentations.

Works Cited

Rubens, Philip. Ed. *Science and Technical Writing*. Henry Holt and Company, 1992.

Schriver, Karen A. *Dynamics in Document Design*. John Wiley & Sons, 1997.

Teacher Resources

Overview and Teaching Strategies

While presentation skills are typically taught in later technical writing classes, once students have learned the foundation of technical writing and have developed their confidence as higher education students, they can be taught in an introductory class.

Ideally, students will have studied introductory topics such as writing style, tone, and layout and prepared other types of written documents before. It is also helpful if students have learned a bit about the profession, in terms of the kinds of duties and tasks that technical writers typically perform and how they might have to produce any kind of content for a technical or scientific organization.

One of the key messages from this chapter is that a presentation is often an extension of another technical writing product and that students need to think carefully about the audience and the goal(s) of the presentation, to ensure the content is tailored to that audience and meets the intended goals. This chapter highlights all the essential steps that students need to follow when planning, designing, and delivering a presentation, with reference to practices that are also important in other types of content development projects.

If presentation skills are to be taught in an introductory course, instructors may wish to reference this essay in the latter half of the course and ask students to develop a presentation that showcases how they approached an earlier assignment (e.g., how they approached a brochure or user manual assignment). In doing so, students will not only develop their presentation skills, they will also have a tangible goal for the presentation. The appendix provides a checklist that students can use when designing, developing, and delivering their own presentations.

Discussion Questions

To engage your students with the ideas discussed in this chapter, consider having them address—individually, in small groups, or as an overall class—the following questions:

1. Thinking back to presentations (including classes) you've attended in the past, which presentations stood out as being effective (or ineffective)? Why do you think they were effective (ineffective)?
2. How do you think a presentation to a live, in-person audience would need to differ from a technical report about the same topic? Elaborate on your answer.

3. Which writing and formatting considerations apply to both written documents and oral presentations?
4. What kind of accessibility issues might you encounter when giving a presentation, assuming a mixed audience? Can you think of any practical ways to minimize those issues?
5. Given the various presentation and interaction styles presented in this chapter, which style(s) would you choose to adopt, and why?
6. What kinds of technology and room issues might you need to consider when planning your presentation? How might you prepare for, and deal with, unexpected glitches?

APPENDIX. CHECKLIST FOR PRESENTATIONS

Phase	Practical considerations	Tick when completed/ considered
Identify the audience and the purpose of the presentation	• Have you identified the purpose and scope of the presentation?	
	• Have you identified the audience profile (e.g., their demographics, motivation, interests, and expectations)?	
Choose presentation tools	• What presentation software (if any) is currently available to you?	
	• What software (if any) do you feel comfortable using?	
	• Would a free trial version suffice?	
	• Which file format(s) do you need to provide?	
Develop the presentation	• Are you going to design your own slide deck or are you required to use a corporate template for your presentation? If so, are there restrictions on the type, color, and graphics that you can use?	
	• Have you taken cultural considerations into account when planning your color scheme?	
	• Have you catered for different accessibility issues when developing the presentation?	

Phase	Practical considerations	Tick when completed/ considered
	• Does your presentation have opening, body, and closing sections? Have you clearly stated the purpose and structure of the presentation in the opening section? Have you summarized the key points of the presentation at the end?	
	• Is the writing style and tone suitable for the target audience? Have you explained any complex jargon if your audience is not familiar with it? Have you adhered to Plain English guidelines?	
	• Have you used tables, charts, diagrams, and/or figures instead of text, when they are required?	
Deliver the presentation	• Have you checked the facilities in the room where you will deliver your presentation? Does your presentation display as intended?	
	• Have you rehearsed your presentation?	
	• Have you identified your preferred presentation style (e.g., formal or informal)?	
	• Do you plan to have handouts (e.g., printouts of your slide deck) and have you printed enough copies in advance?	
	• Have you considered the requirements for any activities that attendees might undertake during your presentation?	
	• If you plan to follow-up with attendees afterwards, have you prepared an attendance list for participants to sign?	
	• If you were unable to answer one or more questions during your presentation, have you followed up with the relevant attendees?	

Phase	Practical considerations	Tick when completed/ considered
	• If you are delivering a virtual presentation, have you familiarized yourself with the software so that you know how to mute/unmute attendees and share/un-share files?	
	• If you are delivering a virtual presentation, have you devised a workable strategy for interaction during the presentation (e.g., should attendees ask questions in live chat only or by putting their hands up)?	

9 Writing Technical Content for Online Spaces

Yvonne Cleary

Until this century, printed manuals were the standard genre for technical documentation.[1] If you bought software (e.g., a word processing program) or hardware (e.g., a computer), you received a hard-copy user manual with information about the product and instructions about how to use it. If you worked as a technical communicator, your job likely involved creating content for print user manuals.

Although print manuals were the conventional and dominant format for product information and instructions for many decades, they have several shortcomings:

- They are expensive to produce. They often run to hundreds of pages, and the associated production, printing, and shipping costs are high.
- Although technical content dates quickly, manuals are slow and expensive to update. This results in a situation where printed manuals are often out of date, refer to a different version, or have inaccurate information. These inaccuracies reduce the credibility of the organization, and, more seriously, lead readers to make errors. You have probably had to use manuals with inaccurate instructions. It is frustrating and can even be dangerous.
- Printed manuals also pose usability problems for readers, who have to navigate long documents, sometimes hundreds of pages, to find answers to questions or learn how to use software, machines, or services.

In sum, printed manuals served a purpose in terms of providing needed instructions and information, but they were cumbersome to develop, produce, update, and use.

---

1 This work is licensed under the Creative Commons Attribution-NonCommercial-NoDerivatives 4.0 International License (CC BY-NC-ND 4.0) and is subject to the Writing Spaces Terms of Use. To view a copy of this license, visit http://creativecommons. org/licenses/by-nc-nd/4.0/, email info@creativecommons.org, or send a letter to Creative Commons, PO Box 1866, Mountain View, CA 94042, USA. To view the Writing Spaces Terms of Use, visit http://writingspaces.org/terms-of-use.

In the past two decades, print manuals have become less prominent, and the term "technical documentation" has given way to the broader concept of "user assistance." User assistance acknowledges that users access a variety of resources on different platforms and devices to learn how to carry out tasks using a product or service. Online user assistance formats include web content, knowledge bases, video, discussion forums, tutorials, e-learning courses, and user interface text. Unlike traditional printed user manuals, this approach enables users to access only the information they need, when they need it, in short, targeted chunks, known as topics. Topics have a limited and specific purpose. They may be delivered on a website as a response to a user query or combined in different ways and delivered in multiple formats. You probably often use topic-based online technical content. For example, if you need to learn how to use a feature of a software program, you might access a website with targeted information that responds to your specific query.

Because various types of user assistance have become more prominent, approaches to developing content have also changed, and many technical communicators now work in teams, creating short-form content, or topics. As a technical communicator, part of your job may involve developing topic-based user assistance content.

Topic-Based Writing: A New Rhetorical Situation

Accessing and using content in manuals is quite different to accessing and using topic-based content. A key difference is that manuals are linear, with an obvious beginning, middle, and end. Consider how you read a book. You have static navigation devices: e.g., page numbers, a table of contents, section headings, and an index. Users access topic-based content, on the other hand, using specific, personalized search queries. Consider how you search for and use information online. To find and navigate information, you rely on very specific contextual cues, like clear titles, short descriptions, and headings, as well as navigation devices like links to related information.

Because users access and use content in different ways in manuals compared to topics, it follows that different processes are involved in planning and developing content for these two formats. For this reason, you need to be familiar with and understand the differences between topics and manuals, the characteristics of topics, and the writing and rhetorical skills required to create content for, and to work in, such contexts.

Of course, print manuals still ship with many types of products, e.g., cars, home appliances, and self-assembly furniture. Developing content for manuals that will be printed or delivered as PDFs continues to be their main job for some technical communicators. Nevertheless, the new context for technical writing in many situations is to develop short-form, topic-based content. The focus of this chapter, therefore, is on the trend towards online, topic-based user assistance, that has replaced manuals for many types of products and services.

Research Basis for Topic-Based User Assistance

Several research developments influenced the shift from long-form linear manuals to modular content in the form of topics. Among them, two were pivotal: Information Mapping®; and minimalism.

Information Mapping® is a system of organizing complex technical information into labelled chunks, developed by Robert E. Horn. It is characterized by information blocks—units of information, and information maps—collections of blocks (Horn 179). Its principles include:

- Categorizing information into types: e.g., procedures or concepts.
- Chunking information into individual blocks.
- Labelling each information block to tell readers what they are reading, and to enable them to scan the content for parts that are relevant to them.

Content designed using Information Mapping® principles is faster to read, easier to process and recall, and results in more accurate task completion (Information Mapping Inc. 11). This system has influenced topic-based writing in several ways, as we will see later in the chapter.

Figure 9.1 shows an example of a document designed using information mapping principles. The document is an information map. Each chunk of content is an information block. The figure shows how this content is easier to scan and to process than long paragraphs of text with no contrast, which is one of the main benefits of information mapping.

John M. Carroll's research on minimalism in the 1980s demonstrated the poor usability of manuals, and advocated task-focused user assistance in minimalist documentation. Carroll rejected the prevailing tendency for instruction manuals to concentrate on explaining a system, instead of helping users to complete tasks using the system. His research was instrumental in the movement towards short-form instruction and topic-based content. These four principles of minimalism (van der Meij and Carroll 21) have influenced the design of instructional materials:

How to design information using an information mapping® approach	
Introduction	This document explains the terms, rules, and advantages of information mapping.
Terms and Definitions	Information mapping uses these key terms.

Term	Definition
Information units	Organizing content into chunks and modules.
Mapping	Formatting content to show relationships and functions of parts of the content.
Information types	Categorizes content according to its function.

Rule 1: Replace prose with structured text	Use information blocks instead of paragraphs.
Rule 2: Replace prose with graphics	Use tables and charts to illustrate data. Use decision graphics and flow charts to illustrate procedures.
Rule 3: Use consistent terms	Avoid unnecessary variation in terminology. Use terms precisely.
Advantages of the method	Enables readers to scan. Enables readers to understand the content. Enables readers to read more quickly. Ensures a common understanding of content. Reduces errors and risks.

Figure 9.1: Content Designed Using Information Mapping Principles.

1. Choose an action-oriented approach. This principle recognizes that people want to get started with a product immediately, and instruction should support users' desire to act.
2. Anchor the tool in the task domain. This principle explains that instruction needs to describe tasks that users can do, rather than features of the system.
3. Support error recognition and recovery. This principle suggests that writers should try to prevent errors and help users to recognize and fix errors they do make.
4. Support reading to do, study, and locate. This principle is concerned with ensuring that instruction does not overload users but helps them to find additional information if they need it.

This four-part approach to minimalism enhances usability by shifting the focus in instructional materials to what people need to do with a system, instead of describing the features of the system. The key characteristic of research on minimalism that influenced topic-based writing was the focus on user tasks. Further, Carroll's work "influenced an IBM technical writer to architect the original DITA [Darwin Information Typing Architecture] schemas for structuring content" (Evia 35). As we will see later in the chapter, DITA is a predominant standard in topic-based writing.

The Emergence of Topics

From the early 1990s, the research of Carroll, Horn, and others led to changes in the design of user assistance. For example, experts including JoAnn Hackos and Ginny Redish explored why structured content with less text is suitable for online readers.

By the 1990s, many software products shipped with online help as well as, or instead of, printed manuals. Online help continued to use a book metaphor, but it introduced the genre of topic. Each "book" contained individual topics (akin to pages in a book) that explained a concept or listed steps in a procedure. Help systems also continued to provide navigation options that corresponded to print manuals, such as a table of contents and an index. They were innovative, however, because they offered additional navigation features that were absent in print manuals: e.g., hyperlinks and search tools. These navigation features have become commonplace in online systems, but they were novel in the 1990s. Users could perform searches of a whole database using a search tool, instead of using an index, or chapter or section titles, to find content. Topics also contained hyperlinks to related topics. Technical communicators wrote topics and created help systems using help authoring tools (HATs), such as *RoboHelp*. For users of the help system—e.g., in a word processing application like *Microsoft Word*—the results were presented in topics, each of which responded to a unique and specific query (e.g., how to copy text, or how to save a document).

The internet has changed how many types of user assistance are created and delivered. Today, instruction manuals are often published online in PDF format, and are less likely to be printed and shipped with a product. A more significant change is that online topics have replaced manuals and online help for many types of products and services, especially for software, applications, and online services.

Recently, we have seen an increasing focus on personalization, the practice of presenting users with the information they need in the moment, and that is specific to their context of use. As we will see later in the chapter, several features of topic-based content enable it to be personalized.

Topic Characteristics

A topic is a unit of information, usually about a single subject. Topics are structured, classified according to type, and designed to be personalized and reusable. Mark Baker used the phrase "every page is page one" (also the title of his book on the subject of topic-based writing) to describe a central characteristic of topics. Each one is self-contained, and when readers encounter a topic, it is outside of the context of any other information on the subject. As writers, we have to work on the assumption that every topic we write may be the first or only information readers encounter: hence, every page is page one (EPPO). We cannot make assumptions about what the reader already knows about the subject. Unlike a manual that a reader is likely to read, or at least perceive, in a sequence, a topic has to function as an independent rhetorical unit.

"Every Page is Page One" (EPPO) Characteristics

Baker described seven characteristics that define topics.

1. **Topics are self-contained.** You will find all the information you need to complete a process or understand a concept in an individual topic.
2. **Topics have a specific and limited purpose** (Baker 88). The content is modular, so each topic can be a building block for a larger document. But a single topic has all the information a reader needs for a specific procedure or concept.
3. **There are different types of topics,** each type with its own characteristics. Each topic has characteristics that identify its type (Baker 97): e.g., procedural topics list steps.
4. **Topics establish their context.** They explain the situations in which they are relevant, either through text (such as titles, headings, and short descriptions) or visuals. This characteristic helps readers to orient themselves.
5. **Topics assume the reader has a certain level of expertise** (Baker 123). They do not explain basic information, only the amount of information needed for the subject at hand.
6. **Topics "stay on one level"** (Baker 130). They do not switch between beginner and expert levels, or abstract and concrete information, for example.
7. **Topics include links.** Although each topic only covers one concept or procedure, abundant hyperlinks enable readers to find out more about related concepts and procedures. This feature makes topics flexible for users.

If you have ever accessed an online support or help center, you have probably used topic-based content.

The best-known system for implementing topic-based writing is DITA. This system is based on XML (eXtensible Markup Language). DITA has many advantages, including its focus on structure and information types, and its potential to enable reusability and personalization. DITA is used in many technical communication workplaces, and you may see job advertisements that mention, or require applicants to know, DITA.

DITA Topics

DITA was originally developed by IBM, and in 2005 it was donated to Oasis, the nonprofit standards body for open-source projects. The Oasis DITA specification defines a topic as "the basic unit of authoring and reuse. All DITA topics have the same basic structure: a title and, optionally, a body of content. Topics can be generic or more specialized" (Oasis "*2.2.1.1 The Topic as the Basic Unit of Information*").

All DITA topics are structured; each piece (or element) of content is labelled, and each element has a specific function. DITA topics usually have all the following elements. Some topics will have many additional elements.

- Topic: every topic in DITA begins with a topic element, written as <topic>. This element contains the topic ID, as well as other attributes (e.g., the product name or the target audience).
- Title: every topic contains a title.
- Short description or abstract: this element provides some context about the content of the topic.
- Prolog: this element contains metadata about the topic.
- Body: this element contains the main topic content, in paragraphs, lists, and so on.
- Related links: topics enable readers to read around a subject, and to get more information if they need it, through abundant hyperlinks.

These elements are key to document creation because they allow you to create and sequence content in ways that are consistent and user-friendly.

Technical communicators use markup to identify parts of a topic in DITA. If you have used HTML, you know that markup tags provide information about content, but the user does not see the markup, only the content. Figure 9.2 shows an example of a DITA task topic. The information enclosed in angle brackets (< . . . >) is markup. This is not part of the content; it identifies elements of the content (e.g. the title, short description, context, and individual steps in the procedure).

```
1   <task id="0001">
2       <title>How to save a file</title>
3       <shortdesc>Whenever you create a file, you need to
    save it to be able to access it later.
4       </shortdesc>
5     <taskbody>
6      <context>File saving is an essential office
    administration procedure. In almost any application you
    use, you will need to save files.
7
8      <steps>
9        <step><cmd>Click on the File menu.</cmd></step>
10       <step><cmd>Click on the Save menu item.</cmd></step>
11       <step><cmd>Type a name for your file.</cmd>
12       <step><cmd>Select a location to save your file, that
    you will be able to access later.</cmd>
13              <info>make a note of the file name and location
    to be sure you can access the file later.</info></step>
14       <step><cmd>Click on Save in the dialog box.</cmd>
15     </steps>
16     <result>Your file is now saved to the location you
    selected and using the file name you typed. </result>
17     <postreq>You can access the file whenever you need to
    use it again<![CDATA[<?xml version="1.0"
    encoding="UTF-8"?>]]>.</postreq>
18     </taskbody>
19   </task>
20
```

Figure 9.2: DITA Topic with Markup.

Topics are combined into larger collections of content using DITA maps. A map is a list of all the topics about a subject, and it describes a particular combination of those topics. It is roughly analogous to a table of contents for a book. Maps "organize topics and other resources into structured collections of information. DITA maps specify hierarchy and the relationships among the topics" (Oasis "2.2.2.1 Definition of DITA Maps").

Although DITA is widely used, it is quite complex and restrictive, and some writers and writing teams prefer to develop topics using other tools. Many alternatives to DITA have emerged. These include Markdown, Lightweight (Lw)DITA, AsciiDoc, DocBook, and schema.org. Some writing teams use wikis to create collections of topics, and some even create topics using word processing applications like *Microsoft Word*.

Information Types

As content users, we expect different types of information to have different characteristics. For example, procedures have steps, recipes have lists of ingredients, forms have fields to be completed, meeting minutes have notes and action points, and so on.

Likewise, structured content is organized according to its type. A feature of DITA, for example, and as its name suggests (Darwin Information Typing Architecture), is that content is categorized by type. Three types of topics are widely used: task; concept; and reference.

- Task topics provide instructions, usually organized in steps, that explain how to do a procedure: e.g., how to set up a social media profile.
- Concept topics explain an idea or give background information, e.g., what a profile is and its purpose.
- Reference topics usually give some kind of extra information, often in tables, e.g., features lists.

These classifications allow the system to ensure that topics are fit for purpose. They "tell the writer how the topic should be written and [tell] the reader how it should be read" (Baker 97).

Single Sourcing, Content Management, and Reusability

You may have heard of the term "single sourcing." This practice involves creating small chunks of content that are centrally stored and can be assembled for use in multiple output formats (Bridgeford 3). This is an important principle in topic-based writing. Topics are written once and stored in a single location, such as a content management system or a database. A technical communicator creates a topic in one source file that can be used in multiple artifacts, e.g. the same content can be used to create instructional topics, knowledge base articles, and website content. Several topics can be combined to produce a longer product, such as a PDF.

A drawback of traditional publishing and printing is that it is difficult to reuse content that is stored in a single, and often sizable, file such as a PDF. Extracting short segments to reuse in other output formats (e.g., on a website) is complicated. The main advantage of single sourcing is that content only needs to be created once, but if the content is managed effectively, it can be reused multiple times.

Reusability is an important goal for organizations because it saves on the costs of developing new content for each version of a product or each device or platform on which the content appears. Effective content management also ensures consistency, because the content only needs to be updated once in the source file, and the updates are reproduced in the output, wherever that content is used. Single sourcing and reusability depend, therefore, on effective content management.

STRUCTURED CONTENT

An important characteristic of topic-based writing is that topics are structured. Evia discusses our "obsession with order and rules" (37) as a long tradition in technical communication, and one that helps readers to process content. Structured content is a method of ordering topic content in a specific way that can be processed by machines. This feature means that it can be reproduced in different ways for different types of user and product. In other words, it can be personalized. Structured content also helps organizations to streamline translation and localization processes (Wachter-Boettcher 155). This is an important consideration when online content needs to be offered in multiple languages.

Structured content has characteristics that users do not need to be aware of, but that technical communicators need to use and understand. One characteristic of structured content is that content is separate from formatting. If you usually develop content using a word processor, like *Microsoft Word*, your document layout and content are integrated. You apply styles (such as headings, bullet points, or captions) to parts of your text as you write. The disadvantage of this approach is that it restricts how you use the content and its constituent parts. It also ties your content to an application. When you copy your text into another application, the formatting may not transfer, or may look different.

To structure content, technical communicators label each element of content to show its function, e.g. heading, body text, caption, list, and so on, but they do not apply formatting (e.g. type size, font, or space) to the element. The content formatting is described in a separate style sheet, and formatting depends on how the content is delivered to the user, e.g. depending on their device and application.

As you can see in Figure 9.2, which is structured using DITA tags, the content is described using tags that explain its structure (e.g., title, steps, result) but the content is not formatted, and the tags do not describe the formatting. The content will look different on different platforms, or depending on how it is combined with other content to build a personalized information product.

User Behavior and Personalization

Think about a recent product you bought or a service you wanted to use online (e.g., setting up a social media account). How did you figure out how to proceed?

As Carroll's research about minimalism showed, users want to explore and get started quickly. They often attempt to use a product before reading instructions. They may try to use even complex applications without accessing user assistance. If they encounter an error or struggle with a task, they may search for an answer online. If the search produces multiple possible answers, they have to select from among the many options, and perhaps deal with information overload.

For these reasons, users are often frustrated. They need user assistance that is clear, concise, targeted, and precise. It must help them to deal with their current situation, whether they are trying to understand a feature or solve an immediate problem. Instead of navigating through a whole website or system, they want to read only the content that is relevant to their query. This is known as personalized content.

Content that is structured and managed facilitates personalization. Topics can be combined in different ways and output formats, for different devices, depending on the user's query and context. When users have a problem or a query (e.g., how do I access a service?, or how do I set up or start using a new product?), they may find assistance on a knowledge base, a self-service portal, a website, or a video platform. All these sources use content developed by technical communicators; in many cases that content is topic-based, and it is often multimodal—a combination of text and images or video. The content is generated for the user, based on the user's query, the device, the mode of delivery, the application, and other user criteria. Effective personalization depends on effective use of metadata.

The Importance of Metadata

You have probably heard the term "metadata" many times. Metadata are data about data, or information about content. The reader does not usually see the metadata, but the information enables the content to be stored, accessed, retrieved, sorted, and delivered to the right user depending on their query. Metadata are, therefore, essential for ensuring that content is reusable and can be personalized. Examples of metadata include:

- File name
- File size
- File location

- Change history
- Audience
- Author name
- Word count
- Keyword
- Tags (e.g., in a DITA file) that define parts of the content (e.g. title, headings, lists, captions, and so on).

If you look at the properties of a document in a word processing program, for example, you will see various types of metadata. Some fields are already populated by the program, and you can add others (e.g., keywords or comments). These metadata will not be visible to the reader, but they contain important information that can identify a file in a database.

Figure 9.3 shows metadata for the *Microsoft Word* document for this chapter, its Summary Properties. Some of these metadata were generated by the program (e.g. the file name). I added others, including keywords and comments. You will not see this information in the final chapter. These are data *about* the chapter.

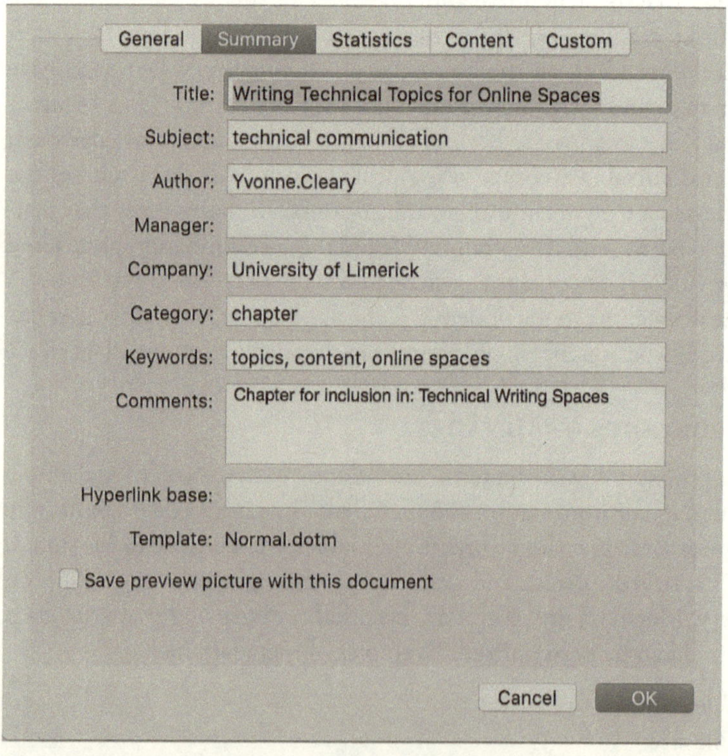

Figure 9.3: Chapter Metadata: Summary Document Properties

Writing Topics: What writing skills do you need?

You need to employ varied skills to write effective topics. Although this type of writing may seem simple, because it is minimalist, in fact, you will need to master many writing skills:

- **Applying technical writing principles.** Many of the writing principles you learn on technical writing courses apply to writing topics. You need to write in a clear, direct, active style that is appropriate for the audience, to ensure your content is clear and understandable. You also need to reduce verbiage and roundabout phrasing. Users need topics that are precise, clear, and concise.
- **Summarizing content.** Writing abstracts/summaries is a technical writing skill that is applied in many writing contexts. For example, reports, journal articles, and e-learning content all contain summaries. In topic-based writing, topics usually include a short description that summarizes the content.
- **Selecting appropriate content.** In any rhetorical situation, you have to develop the skill of choosing content that is relevant, helpful, and appropriate. The principle holds in topic-based writing. Users do not want to be patronized with information that is too basic, or confused by information that is too advanced. Your topics must be purposeful and helpful. Even though topics are usually short, decisions about what content to include, and what to omit, require skill that comes with practice and experience.
- **Taking a disciplined approach to writing.** While all writing requires discipline, this principle is essential in topic-based writing. As we have seen, an important characteristic of topics is that their purpose is limited to answering one question. To ensure that you focus only on the topic content, it is helpful to write one topic at a time, and to avoid thinking about the topic as part of a larger body of information. This principle is captured in Baker's phrase, 'every page is page one.'
- **Chunking information.** Chunks are blocks of information. We saw in the Information Mapping approach how chunks can help users to scan and process information more quickly. Because topic-based writing is modular, rather than linear, chunking is an essential skill.
- **Applying other principles of information mapping.** As well as chunking information into blocks, other information mapping principles help you to write focused topics. Labelling content clearly ensures readers can process it effectively. In topic-based writing,

labels include the topic title, headings, and short descriptions. Categorizing information into types, each with its own characteristics ensures that readers know how to use topics. As we have seen, the three most common information types are task, concept, and reference. Each of these topic types has its own rhetorical characteristics. For example, task topics are written as steps, concept topics are written as paragraphs, and reference topics typically include tables.

- **Applying principles of minimalism.** A central principle of minimalism is its task orientation. Focusing on tasks and actions in your topics helps readers to avoid making, and to recover from, errors.
- **Writing metadata.** A skill that technical communicators increasingly require is the ability to write good metadata. Even though your readers will not see the metadata, effectively labeling your content ensures that it is reusable, consistent, searchable, scalable, and responsive.

Additional Skills for Writing Topics

Several skills in addition to writing are becoming essential in many technical communication workplaces, as technical communicators move from writing long-form documents to shorter, topic-based content. As more organizations move user assistance online and self-service portals become more commonplace, demand for these skills will increase further.

Collaboration

Sole technical communicators were responsible for the design and delivery of individual parts of manuals, or even entire manuals, in the past. This model of content development has become known as the craft tradition of technical communication (Bridgeford 7). As manuals have been replaced by user assistance topics for many products and services, you are much more likely to work with teammates to develop collections of topics, that are stored in content management systems. You may be assigned a set of topics that you have to create, track, and update individually. Nevertheless, you are likely to work with developers to ensure the content is correct, with graphic designers to select appropriate visuals, and with other writers to develop standards, style guidelines, and review processes. Teams are often interdisciplinary and virtual, with members located in different sites and possibly even different countries.

For all these reasons, collaborating effectively within interdisciplinary teams to produce useful and consistent topics is an essential skill. Tracy

Bridgeford describes the skills required for true collaboration in a content management environment as "networked agency" (8–10).

Using Tools for Topic-Based Writing

The technologies we have discussed so far are tools, for they are human creations we use to achieve a specific objective—much in the same way we use a hammer to pound a nail or a saw to cut a piece of wood. In addition to these tools, we need to use applications like word processors and text editors. In industry, these applications are also known as tools. Although topic-based writing is now commonplace, there is no industry standard application or approach that you need to learn. Depending on the organization you work in, you may use one or more of the following approaches:

- Create topics using a word processor.
- Create topics in a wiki.
- Create topics using the DITA open toolkit.
- Create topics using lightweight markup languages such as Markdown or text document formats like AsciiDoc.
- Use DITA editors such as Oxygen XML Editor.
- Use applications such as MadCap Flare or FrameMaker, or in-house editors.
- Use content repositories, such as a database or content management system.

Indeed, you may combine more than one of these approaches, or use additional approaches not listed here to develop topics in industry.

Trends in Topic-Based Online User Assistance

Approaches to technical communication are continually evolving because this is a dynamic profession that is heavily influenced by developments in technology. For that reason, you will need to be prepared to learn new tools, applications, and strategies all the time. These are some current trends in topic-based writing that are likely to influence how you work in the future.

Intelligent Content

The term "intelligent content" describes content that is "structurally-rich and semantically-categorized, and is therefore automatically discoverable, reusable, reconfigurable, and adaptable" (Rockley and Gollner 33). The more adaptable and responsive content is to a user's specific query, the more "intelligent" it is. According to Carlos Evia, "[topic-based] writing is the basis of several technical communication and intelligent content practices" (9). We have discussed many of these practices in this chapter:

- Developing reusable content.
- Using metadata.
- Structuring content.
- Identifying suitable applications for content development and management.

Combining these practices facilitates the development of intelligent content, now a goal in many organizations.

Microcontent

Topics may be the dominant technical communication output format for now, but even shorter formats are becoming commonplace. Microcontent comprises short snippets of content that are embedded in a device or application. This content may be as short as a word or a phrase. Ari Hoffman explained contexts where users use microcontent, and why writers need to prepare to write in this way:

> Printed user manuals moved onto a desktop screen, the desktop turned into a laptop, the laptop into a smartphone, then into a smartwatch, washing machine, VR headset, or even an Alexa on your customer's night stand. We need to stop "writing manuals" and start crafting usable, findable micro-content. (7)

The Information 4.0 Consortium describes the future of content as "molecular" rather than modular.

User- or AI-Generated Content

A trend that you have probably noticed, and perhaps even contributed to, is the movement towards user-generated content. Users are adept at solving problems, and when we solve problems, we often like to share our solutions, either through discussion forums, videos, or other formats. This content is then available for other users. You may have used user videos on YouTube or discussion forums online to understand a procedure. If so, you have used user-generated content.

Although user-generated content is often accurate and accomplished, sometimes it provides incorrect information, or content that discredits the product or organization. For these reasons, many organizations employ technical communicators to moderate or curate user-generated content. In such a role, your job is to ensure that the content is correct and reflects well on the organization. You may also incorporate user-generated content into professional user assistance topics.

Another recent trend is towards automatically generated content using artificial intelligence systems. Topics developed using generative AI may have many inaccuracies, and your role may be to test and revise these topics.

Conclusion

In contemporary work contexts, technical communicators are likely to work in teams, and to write topics, or short chunks of content, that may be combined to generate information products. Topics have several characteristics that distinguish them from other forms of content; in particular, they exist as independent rhetorical units. The best-known topic-based writing system is DITA, but technical communicators use many approaches to develop topic-based content. Likewise, many skills are required, that fall into three broad categories: writing, collaboration, and using tools. Because technical communication is a dynamic discipline, topic-based approaches to writing will continue to evolve into the future. It is important for you to be aware of the state of the art and to keep up with trends and practices.

Works Cited

Baker, Mark. *Every Page Is Page One: Topic-Based Writing for Technical Communication and the Web*. Laguna Hills, CA: XML Press, 2013.

Bridgeford, Tracy. *Teaching Content Management in Technical and Professional Communication*. Edited by Tracey Bridgeford, Routledge, 2020.

Carroll, John M. *The Nurnberg Funnel: Designing Minimalist Instruction for Practical Computer Skill*. MIT Press, 1990.

Evia, Carlos. *Creating Intelligent Content with Lightweight DITA*. Routledge, 2019.

Hoffman, Ari. "The Evolution of Tech Comm: Directing the Content Experience." *Intercom*, Jan. 2018, pp. 6–8.

Horn, Robert E. "Results with Structured Writing Using the Information Mapping Writing Service Standards." *Designing Usable Texts*, edited by Thomas M. Duffy and Robert Waller, Academic Press, 1985, pp. 179–212.

Information 4.0 Consortium. "*What is information 4.0?*" https://information4zero.org.

Information Mapping Inc. "*The Information Mapping Method: 30 Years of Research*." 1999, http://www.writec.com/Media_public/IMI_history_and_results.pdf.

Oasis. "*2.2.1.1 The Topic as the Basic Unit of Information*." 19 June 2018, https://tinyurl.com/wyx5znsb.

Oasis. "2.2.2.1 Definition of DITA Maps." 17 Dec. 2015. https://docs.oasis-open.org/dita/dita/v1.3/os/part1-base/archSpec/base/definition-of-ditamaps.html.

Rockley, Ann, and Joe Gollner. "An Intelligent Content Strategy for the Enterprise." *Bulletin of the American Society for Information Science and Technology*, vol. 37, no. 2, 2011, pp. 33–39.

Van der Meij, Hans, and John M. Carroll. "Principles and Heuristics for Designing Minimalist Instruction." *Minimalism Beyond the Nurnberg Funnel*, edited by J.M Carroll, MIT Press, 1998, pp. 19-54.

Wachter-Boettcher, Sara. *Content Everywhere: Strategy and Structure for Future-Ready Content*. Rosenfeld Media, 2012.

Teacher Resources

Overview of Teaching Strategies

Topic-based writing for online delivery is a paradigm shift. Most students, most teachers, too, are familiar with writing long documents, and with solo writing projects. Technical content is already perceived as difficult. Add to that the many new terms and concepts in topic-based writing. For all these reasons, this subject might seem intimidating to students. It's a subject that we as instructors may struggle with too.

I teach this content through various activities that I outline in more detail in this chapter: Cleary, Yvonne. "Teaching Topic-Based Writing." *Teaching Content Management in Technical and Professional Communication*, edited by Tracey Bridgeford, Routledge, 2020.

I find it helpful to introduce students to the idea of topics by asking them to explore professional topics that are available online (see Activity 1). This activity helps them to recognize that they are already familiar with this genre. They have probably used topics multiple times. When they are familiar with the features of the genre, I ask them to work on Activity 2.

Activity 1: Explore the Features of Professional, Published Topics

Find examples of online topics (e.g., in the Amazon Knowledge Center, the YouTube Help center, or the Broadcom Knowledge Base).

1. Ask students to discuss the ways in which a topic matches Baker's topic characteristics. In which ways does it differ from those characteristics?
2. Compare and contrast topics from different organizations. Look at headings, subheadings, summaries, short descriptions, body text, and links, for example.

3. Look at the source code for a topic (e.g., in Google Chrome, you can view the source code by clicking on View/Developer/View Source). In the source code, find examples of metadata.

Activity 2: Develop a Topic

1. Provide a list of example topics (e.g., for common tasks or concepts that students will understand or be familiar with).
   - Assign each student an individual topic to work on. Ask them to write a topic. They should include metadata and label parts of the content in ways outlined in this chapter. It is not necessary to use DITA or to use specialist software. They can use Microsoft Word or a similar word processing tool. An alternative is to use a wiki, where each student develops content for one page of the wiki. This approach enables students to collaborate on a large body of content.
   - Discuss in class how the topics could be combined in different ways to create longer documents.
   - Discuss in class how this type of writing differs from writing essays or longer documents.

Discussion Questions

1. Based on your understanding of this chapter, explain how topics differ from other technical writing formats, such as manuals. In what ways are topics more useful and usable than printed manuals?
2. Find an example of an online topic. Identify the topic type (task, concept, or reference). What are the features of this topic? Consider, for example, headings, short descriptions, formatting, visuals, and links.
3. In a small group, discuss each of Baker's seven Every Page is Page One characteristics. Which of these characteristics would you find easiest to apply in your writing? Which ones have the most impact for readers?
4. Consider the skills needed for topic-based writing that are discussed in this chapter. Which of these skills have you developed? Which ones would you like to develop further? What strategies or activities would help you to develop these skills?

10 Social Media as a Space for Today's Technical Communication Work

Clinton R. Lanier

The way we communicate has dramatically changed over the past two decades.[1] Not long ago a college student might have taken a break between classes by reading a magazine or talking to friends. Today, when 99% of college students own smartphones (Seilhamer et al.), they might instead view the latest posts on Instagram or TikTok, and they would not be alone in their browsing. In fact, according to a study by Pew Research, 72% of adults in the U.S. are active social media users. With so many people using Facebook, Snapchat, and other social media networks, it is no wonder that they are also being used for less "social" communication as well (see, for example, Bowden and Charles-Smith et al.). Even the field of technical communication has found a way to use social media, and that is what this chapter is going to focus on.

Understanding how technical communication and social media relate is an important topic according to past research. In one study, technical writers were asked what the most important trends or technologies were for the field of technical communication. One of the common answers was "social media," and specifically Facebook, Twitter (the platform now knows as X), Instagram, and YouTube (Lanier). In another study, researchers looked at job postings for technical communicators to understand the types of jobs they were getting and what skills they needed for them. One of the biggest job sectors was in "social media writing," and one of the skills required for many of the technical writing jobs included an understanding of social media (Brumberger and Lauer).

This chapter highlights the relationship between social media and technical communication. First, we will define social media so that we have a

---

1 This work is licensed under the Creative Commons Attribution-NonCommercial-NoDerivatives 4.0 International License (CC BY-NC-ND 4.0) and is subject to the Writing Spaces Terms of Use. To view a copy of this license, visit http://creativecommons.org/licenses/by-nc-nd/4.0/, email info@creativecommons.org, or send a letter to Creative Commons, PO Box 1866, Mountain View, CA 94042, USA. To view the Writing Spaces Terms of Use, visit http://writingspaces.org/terms-of-use.

better understanding of what we mean when talk about it. Then we will cover some of the main ways in which social media is used by technical communicators, looking at examples of each used in the real world to help technical writers perform their jobs. By the end of the chapter, you will understand what tasks you may use social media for in the workplace, and how those tasks are performed.

What is Social Media?

When people hear the term "social media" they often think of the platforms we use to connect to friends and family, like Facebook, Instagram, and YouTube. But social media means more than that. In this section we want to try to better understand the meaning of the term social media, so that we can better talk about it in relation to technical communication.

One of the best definitions of social media comes from an article by Carr and Hayes, which states:

> Social media are internet-based channels that allow users to opportunistically interact and selectively self-present, either in real-time or asynchronously, with both broad and narrow audiences who derive value from user-generated content and the perception of interaction with others. (Carr and Hayes)

The definition does not name any specific platform, but instead focuses on the functions they have in common and what users do on them. First, these platforms allow users to share content of all kinds, like text-based messages, images, and videos. This content is considered "user-generated," meaning that the social media users are responsible for creating and distributing it. The opposite might be "corporation-generated," content that is made by a newspaper or publishing company. This is an important distinction because it means that users can create and distribute whatever content they would like without having to ask permission to post it. It also means the content we receive from social media channels can come from almost anywhere and be created by almost anyone—creating concerns on how accurate and credible information might be.

Social media also allows users to define the audience with whom they share their content, and that audience can be as large or as small as they choose (based on the settings of the user). A typical website, on the contrary, is either public or private, with no in-between. If you were to put something on a webpage, you have little control over who sees it. It might be seen by people who are not your target audience, or people who you would rather not see it.

People can also interact with each other and the information shared through social media. They can ask questions, argue, give feedback, or provide corrections. And they can do this instantly in real time, as Facebook Messenger demonstrates, or sometime in the future, like days, months, or even years later (e.g., responding to or forwarding a Tweet that went out weeks ago). Conversely, digital media platforms like webpages have little ability to provide interactions, and any they do provide (like comments after an article on a page), are limited to asynchronous (i.e., time delayed) interaction with no ability to have real-time conversations. Still other platforms, like forums, are great at interactivity, but fall well short of social media's ability to feature multi-media (like images or videos), or to broadcast messages (like questions or information) to networks.

So social media is more than the platforms we use, it is what they allow us to do that makes them social media, and what they allow us to do is communicate in ways that were not possible even two decades ago. And while certain platforms specialize in certain types of media (for example, YouTube specializes in videos and Instagram specializes in images), they all share the same features in common. It is these features that technical writers can harness to create better information products for end users. The next section discusses how social media is being used in technical communication, and expands on the idea that through these platforms, technical communicators can do their jobs even better.

How is Social Media Used in TC?

Now that we have defined social media, we will turn to understanding how it is being used every day in technical communication. While this section cannot cover every major way in which it is used, it will cover some of the more important areas where social media can help technical writers create better information products for users.

Documentation Distribution

When people think of technical communication, one of the first things that might come to mind is technical documentation. In this case, documentation means information that teaches or explains to readers how to use or do something. Examples of the kinds of documentation you, as a technical writer, might create include instruction manuals, how-to guides, and reference books. Technical writers have been creating such technical documents for decades and, in fact, the technical communication profession itself was formed to better study and create such products. These documents were the stock-in-trade of the technical writer, who worked diligently for years to help people better understand how to use their technology product, be it a small software application or a military jet fighter.

For half a century this work was carried out first on typewriters and then on computers. The technical documentation—the instruction manuals and guides—were always intended to be printed. Sometimes the instructions were little more than a single sheet of paper, folded up and included with the packaging of a product (which you can still find today with many small or simple products), and other times they were thick books, bound together on pages with color illustrations. The point is, these *information products* (as items that technical communicators make are often called) were always printed out, always delivered with a specific product, and always meant to be held in the hands of the user.

But over the past two decades, technology has dramatically changed how these staples of technical communication are made and delivered. Not only have specialized software programs helped technical writers create information products more quickly and easily, but they have also transformed the way people read and use them, and little has transformed them more than social media. While *printed* documentation will always have a place in technical communication, more and more instructions and "documentation" are being produced via video, which is then distributed through different platforms, most especially through YouTube.

YouTube is the second most popular social media platform in the United States (behind Facebook) and the most popular video streaming platform in the world. According to the company's own statistics, over 1 billion hours of video are watched every day (YouTube for Press). And while many of those are fun and entertaining, a good number of them are also videos that provide instruction to users. In fact, there really is not a process or procedure you can think of, whether it is building a model airplane or flying a real one, that does not already have at least a handful of videos on YouTube to tell you how to do it.

And that goes for very complicated, very technical products as well. Tutorials and guides for using complex technology are often uploaded to YouTube by their manufacturers. Adobe, one of the largest technology companies in the world and maker of applications like the Portable

Document Format (PDF), Flash, and Photoshop, created dedicated YouTube channels to help users better understand their products.

The Adobe-run channel, *Adobe Photoshop*, features hundreds of videos explaining the intricacies of their popular graphic editing software. Photoshop is an extremely sophisticated application with a wealth of features and functions. Each feature has multiple settings and can be combined with other features, each with their own sets of modifications and changes. The result is a computer program with thousands of potential ways to be used. Imagine designing a user's manual for that? Adobe embraced social media because they knew that if their users wanted to understand how to truly use their product, they had to move instruction from books to videos, and YouTube allowed them to share those videos with millions (though they also publish numerous books and online articles to help their users as well).

It is not just large companies using YouTube to provide technical documentation, small companies use it too. GeoInsight makes a single product used for obtaining water samples for environmental testing. Though the device is relatively simple, it comes in different designs and there are a number of ways to use it. While the company keeps sets of written documentation that can be downloaded from its website, it also created a channel on YouTube called *Hydrasleeve* to host how-to videos, demonstrations, and studies. It serves a very small population of users, but it provides a different method through which the company can give those users the information they want.

Past studies have shown that there are right and wrong ways to create technical videos for YouTube. Instructional videos should take the time to introduce topics and tell the users what they will be able to do by watching. They should also demonstrate content and both explain and perform the steps (instead of merely performing them) (Swartz). While performing the process or steps, the videos should have a high production quality and use static images as well as video and include textual messages to reinforce the images. Finally, they should have background music but less background noise, and include a fast speaking rate (Hove and van der Meij).

The use of videos, through social media platforms like YouTube, will surely increase as internet speeds become even faster and more people carry devices on which to watch them. They are a convenient and effective way of supporting users of technology.

Audience Analysis

Social media is also really useful for understanding an audience. Though technical communication is best known for the products it creates, like instruction manuals or how-to videos, much of a technical communicator's time is spent in the planning process. And within that process, one of the most important stages is understanding the people who will actually be reading the instructions or watching the videos. Without this knowledge, technical communicators may have no idea what or how much information their readers actually need. If a product's documentation does not match a reader's expectations, there could be some very negative results, like bad product reviews or a loss in sales. To make sure that the information they are providing is actually the information the readers need, and to also make sure they are giving the information in the right way, technical communicators will study their audience through a method often referred to as an "audience analysis."

Audience analyses may be carried out in a number of different ways, but they all have the same goal: to understand the document's readers (Ross). Among the many aspects about the audience that technical communicators want to know include how much the readers already know about the product or technology and what they use that product or technology for and how, where the readers are from and what their primary language is, and what their ages are and how they learn new information. These characteristics and many more are all combined so that technical communicators can build a comprehensive picture of what the reader looks like, which is often called an "audience persona." This audience persona is an imagined person (or people, as technical communicators may build 5-7 or more of them) that represents the reader of the documents being created (Lam and Hannah). This helps the technical communicator understand exactly what information should be included in the manuals and how it should be presented.

Conducting audience analyses and creating personas used to be very difficult, time consuming, and expensive (Hovde). Technical communicators had to use methods like surveys, interviews, and focus groups to find out information about their readers. Researchers would send paper-based surveys in the mail or through email, both asking customers about themselves. Unfortunately, this method requires you to already know who your readers are before you conduct the research. Another problem is that rarely do people respond to surveys. In fact, surveys by mail usually only receive about a 50% response rate (Keeter et al.), while email surveys only get about a 24.8% response (Fluidsurveys). Interviews and focus groups present challenges because they require participants to give up their time and provide a lot of information that they might not feel comfortable giving to strangers.

Social media, though, makes the process of finding out about an audience far more convenient (Humphrey). Platforms like Facebook, Twitter (X), and Instagram help technical writers better understand their product's customers, and can help writers identify important characteristics about them, like where they live, how old they are, whether they have pets, and even what events or news stories they are following or interested in. Information like this and more helps the writer then draft documentation or make videos that better fit the audience, whether by informing writers to use examples and metaphors better understood by people of a certain age, or to use images and colors that are more culturally appropriate. It is not that this information could not be attained without social media, it is that social media makes the process of information gathering much easier.

Twitter (X) and Instagram, because they allow people to see what their users post (as long as their profiles are marked 'public'), can help technical writers build audience personas based on what their readers are talking about (Humphrey). Perhaps the easiest method of using them is by conducting a simple search for certain keywords. For example, if a technical communicator was designing an instruction manual for a SQL database, they might conduct a keyword search for "SQL" or "SQL database." From the search they might see dozens or even hundreds of users who have posts containing the word, "SQL." The list of posts derived from this search could then be narrowed down to the most relevant posts (maybe people talking about using SQL databases). Finally, from that list, technical communicators can look at the profiles of the people who posted them and gain important insights about them, like where they live and what they do. From all this information, a full picture of the potential audience for an instruction manual for a SQL database can be formed, and accurate reader personas can be created.

Facebook provides even more insight about its users, which can also really help technical communicators get to know their own audiences. The first step, as in the case of Twitter (X) and Instagram, is to find the relevant users of the products the technical writer is writing about (Humphrey). Because Facebook includes among its features Pages and Groups for users interested in specific interests, technical writers may need only to find the Pages or Groups related to what they are writing about, and then begin gathering information about the group members and people who like the pages. As in the case of Twitter (X) and Instagram, details like where people live or what they do can be easily found (depending on the users' privacy settings), but Facebook can provide much more than that. An audience analysis using Facebook as a source might also tell us where people work, what education level they have reached, what their primary languages may be, and even whether or not they like dogs (though that may not always be relevant).

Now, understandably, some might feel a bit creepy looking at all of these users and collecting their information. If that is the case, then these platforms also permit other great ways of conducting reader research. All three of them allow the creating and posting of polls, so technical writers can ask users questions about themselves. These polls can either be included as page posts (or Instagram posts or a tweet on Twitter (X)), or as advertisements to platform-users who have been identified as potential audience members. Whatever method is chosen, all of the information helps technical communicators create better information products that will be more effective and helpful for the readers.

Reader Feedback

It is sometimes very difficult to understand whether the information products that technical communicators create are actually effective or not. Is the information in them correct? Is it easy to understand? Does it actually help the readers accomplish their tasks? Aside from conducting a thorough analysis of the audience, as explained in the previous section, how can we find out this information? The easiest way is to get feedback from the readers themselves (Kahn). If we know, for example, that certain steps in an instruction manual are confusing, then we can fix them for the next readers. But unfortunately, getting that information is not at all easy.

Years ago, the only method for getting reader feedback was by phone or email. At the back of most any manual would be found an email address where people could send general inquiries about anything relating to the product, like questions about defects or the product's use. This email address was also where readers could provide feedback about the documentation itself. Once someone sent an email, that feedback would eventually find its way to the technical writers who created the manual, but it could take some time for this to happen (feedback was rarely, if ever, immediate). And if the technical writers needed to clarify the information received, they would have to email the reader back, hoping that the person would then expand on their previous comments. Ultimately, getting information from readers that provided insight about documentation was difficult to do, and so it was rarely pursued.

Today, however, getting readers to provide information about products or technical documentation is far easier thanks to social media (Astoria Software). Social media platforms, especially Facebook, Twitter (X), and

LinkedIn, allow technical writing teams to gather feedback from individual readers or whole audiences. If the company or product has its own social media channels—like a Facebook Page or Twitter (X) profile—then these can easily be used to request this type of feedback. A post could simply ask readers about their impressions about the product's documentation, interface, or support system, and then note the responses that follow (Kahn). The amount of feedback received will no doubt vary based on how many Twitter (X) followers or Page Likes there are, but even a minimal amount of feedback can alert technical writers to issues or problems with their information products that they had not seen themselves.

Perhaps just as important is the ability of technical writers to more quickly communicate with the readers that actually provide the feedback. In the past, it could take days or maybe weeks for the comments from readers to reach the desks of the technical writers the comments were meant for. But now, the information, comments, or questions that readers post to social media platforms can be responded to instantly (Carr). Technical communicators can ask follow-up questions to clarify or otherwise get a better sense of the readers' concerns. They can also, in turn, address the issue the reader communicated about, and then send it to the reader to make sure the problem that was identified was actually fixed. This type of rapid, back and forth feedback can help advance the usability of a product's documentation immensely, because if a single reader has a problem with the information, then there is a good chance other readers will have that problem too.

In addition to gathering unsolicited feedback from readers, technical communicators can also use these same social media platforms to test out new information products and request information about them (Gearhart). For example, a technical writing group can create a post that features a new software interface on the product's Facebook Page, and then ask the product's users what they think of it. Is the interface better or worse than the old one? Is the information on it easy to find or is it cluttered and confusing? The information they receive can help the writers make changes to the interface (or instruction manual, etc.), and create a better product. This type of usability testing always existed, but it is much easier to carry out now because of the reach of social media platforms.

In the previously noted case, the social media platforms' advertising ability would also be very convenient. If a product did not have a large social media following, but the product's writers still wanted to gather feedback about the documentation, interface or other information product, they can create advertisements that would reach the right people. Using strategies from the second section (conducting an audience analysis), advertisements could be aimed at the right type of social media user, people who might use the product being designed. Though this method costs money, it is still a much easier, and potentially far less expensive, manner of gaining insight into the effectiveness of the documentation being written.

TIPS FOR USING SOCIAL MEDIA IN TC

This section provides some tips for actually using social media to carry out technical communication. This is not at all as easy as you might think. As with communicating through any other media, there are a lot of things to consider before you begin creating a Facebook post or YouTube video. Three specific questions to ask are: When should social media be used instead of other media? What platform might be best to use? How should technical information be presented on them? The following discussions will try to answer these questions in turn.

WHEN SHOULD SOCIAL MEDIA BE USED?

Where technical writing is concerned, what you use to communicate to an audience will depend on the context of the writing situation. If your audience is working in a circumstance where they will not have access to the internet, then perhaps the information they need is best communicated offline, as in a traditional hard copy manual. But if you are sure they will have access to the internet and the ability to stream multi-media, then making information available through social media becomes an option.

Another item to consider is the amount of technical information to be communicated. Social media does not lend itself to large amounts of information with multiple sections, reference items or explanations. According to research conducted by the software company, TechSmith Corporation, 52% of YouTube viewers prefer instructional videos no longer than 6 minutes (TechSmith). And while both Facebook and LinkedIn allow users to write longer posts, they have no way to maneuver or navigate within them (like linked tables of contents or ways to jump to different subjects), forcing readers to scroll down and scan for information needed. Finally, does your audience actually use social media? Each social media platform has specific audience demographics that are easy to identify (for example, see Chen), and if the audience you are trying to reach does not fall into a group using the platform you are considering, then you should find another way to reach them, maybe one that does not even include social media.

WHAT PLATFORM WOULD BE BEST TO USE?

Again, the platform you use to communicate to audiences or use to research your audience or gain feedback will depend on your goals and audience. If you hope to conduct research about your readers to create better documentation for them, then using a platform that allows you to conduct such research would be ideal. Platforms like Twitter (X) or Instagram let you see details about users if their accounts are public, whereas Facebook and LinkedIn may not provide any details at all. However, if you hope to share detailed information with those readers, then you want to choose a platform that allows you to provide longer forms of information, like Facebook, LinkedIn, and YouTube.

Likewise, if your goal is to share multi-media, like an instructional video, then choose those platforms that allow you to distribute that type of media, especially YouTube, which was designed to broadcast video. Conversely, while YouTube is perhaps the best at distributing videos, it is far less effective at communicating directly to audiences. It does have a commenting feature, but the near real-time ability to converse with customers or product users that Twitter (X) or Facebook offer is much better. So, before deciding which platform to use, try to understand what you want to achieve and who you are trying to communicate with, and then let that information guide your decision.

HOW SHOULD INFORMATION BE PRESENTED?

Finally, you need to understand how best to present information on these various platforms. Because each one is so different, there are different methods and best practices for each, and so a full accounting of how to do it is out of the scope of this chapter. But there are some general rules that can be followed which may help, the most important might be to remember that the information should be brief, professional, interactive, and multi-media. First, remember that social media is not the best platform for long form information (like an installation guide with hundreds of steps). Instead, present information that can be shared in small "bites," like quick tutorials, explanations, or topics. Alternatively, you can use the post to direct readers to longer forms of information if needed.

Also, ensure that your posts are professionally written. Grammar, spelling, tone, and voice should all reflect the same level expected in other forms of a product's documentation. The type of casual language (not to mention emoticons or funny memes and gifs) you might use when posting on your personal accounts for friends or family to see would be out of place when presenting the sort of precise information technical writers are expected to relay. An attempt to be entertaining could distract from an important message needed by your readers.

Next, understand that information on social media platforms is interactive. Readers can ask questions about it, comment on it, rate and review it, share it, and copy and change it. Unlike a PDF instruction manual that might be posted on a company's website and then forgotten about, the information on social media platforms could continue "living" for ages afterwards. So, be sure to respond to the interactions; to the questions, comments, ratings, reviews, and changes. Finally, take advantage of the abilities of the platform you use. Each of them allows you to share multi-media, which may make it easier for an audience to understand the information presented. Things like animated GIFS, videos, and different types of images can be easily integrated into posts and tweets, so find opportunities to use them for your audience.

Conclusion

Social media is certainly not a passing fad. Though the specific platforms may come and go, there will most likely always be apps that allow people to connect to each other to share content (a very basic, but still very relevant, definition of social media). Some of the biggest platforms continue to remain popular among all age groups and demographics, especially Facebook and YouTube, two of the oldest. And as they continue to be used by so many people, they will also continue to be an excellent space for technical communicators to carry out their work. Simply put, technical communicators go where their audience is. If their audience prefers printed manuals, then that is what writers will make for them. At the moment, much of the audience is spending their time on Facebook, Instagram, Twitter (X), and YouTube, and so harnessing those technologies for technical communication makes a lot of sense.

"Documentation" is being redefined as technical writers use social media to distribute content, especially video-based content. Gone are the days when someone who purchased software had to rely on a dusty manual to learn about it. Now, quick video tutorials about products and their processes, uploaded by the organizations that made the products, are becoming increasingly common, giving audiences new modes for learning. Much of this content is being designed and produced by technical communicators.

The relationship between the audience and the technical writer is also changing thanks to social media. It is easier to find out and understand who their readers are now. With a simple search on Facebook or elsewhere, technical communicators can construct a comprehensive picture about who their audience is, what kind of information they need, and how best to present that information.

Finally, social media platforms are placing technical writers and their readers into a much more collaborative relationship. While once they were separated by time and space, and the only way for a reader to contact a writer was through a vague email address at the back of a manual, they can now converse with each other through multiple platforms, like Twitter (X) or Facebook. Readers can point out issues or ask questions about the process of using a product, and technical communicators can ask for feedback about everything from new interfaces to a set of instructions. This exchange of information could lead to better information products that benefit all of its users, not just those the technical writer has been talking to.

The list of ways that social media is changing and helping technical communication is by no means exhaustive, there have been a wealth of studies that show how social media platforms are being used in technical communication in many different ways. And as time passes and more features and functions are added to social media platforms, there is no doubt that they will be found useful in even more ways. And the technical writers of the near future will need to be well versed in using social media, not just for keeping in touch with their friends and family, but for carrying out technical communication as well.

Works Cited

Bowden, Melody A. "Tweeting an Ethos: Emergency Messaging, Social Media, and Teaching Technical Communication." *Technical Communication Quarterly*, vol. 23, no. 1, 2014, pp. 35-54.

Brumberger, Eva and Claire Lauer. "The Evolution of Technical Communication: An Analysis of Industry Job Postings." *Technical Communication*, vol. 62, no. 4, 2015, pp. 224-243.

Carr, Caleb T. and Rebecca A. Hayes. "Social Media: Defining, Developing, and Divining." *Atlantic Journal of Communication*, vol. 23, no. 1, 2015, pp. 46-65.

Carr, David F. "How Social Media Changes Technical Communication." *InformationWeek*, 2012, http://www.informationweek.com/how-social-media-changes-technical-communication/d/d-id/1102043?

Charles-Smith, Lauren E., et al. "Using Social Media for Actionable Disease Surveillance and Outbreak Management: A Systematic Literature Review." *PloS ONE*, vol. 10, no. 10, 2015, https://doi.org/10.1371/journal.pone.0139701.

Chen, Jenn. "Social media demographics to inform your brand's strategy in 2020." *Sprout Social*, 2020, https://tinyurl.com/449hzves.

Gearhart, Bill. "Incorporating Social Media into a Technical Content Strategy White Paper." *The Center for Information-Development Management,* 2014, https://tinyurl.com/mptxv4p6.

Hovde, Marjorie R."Tactics for Building Images of Audience in Organizational Contexts: An Ethnographic Study of Technical Communicators." *Journal of Business and Technical Communication, vol.* 14, no. 4, 2000, pp. 395-444.

Humphrey, Aaron. "User Personas and Social Media Profiles." *Persona Studies,* vol. 3, no. 2, 2017, pp. 13-20.

Kahn, Sarah. "Harnessing Social Media for Rapid Usability Testing." *Envato Tuts+,* 2012, https://webdesign.tutsplus.com/articles/harnessing-social-media-for-rapid-usability-testing--webdesign-8907.

Keeter, Scott, et al. "What Low Response Rate Mean for Telephone Surveys." *Pew Research Center,* 2017, https://www.pewresearch.org/methods/2017/05/15/what-low-response-rates-mean-for-telephone-surveys/.

Lam, Chris, and Mark A. Hannah. "Flipping the Audience Script: An Activity that Integrates Research and Audience Analysis." *Business and Professional Communication Quarterly, vol.* 79, no. 1, 2016, pp. 28-53.

"Response rate statistics for online surveys – what numbers should you be aiming for?" *Fluid Surveys University,* 2014, https://fluidsurveys.com/university/response-rate-statistics-online-surveys-aiming/.

Ross, Derek, G. "Deep Audience Analysis: A Proposed Method for Analyzing Audiences for Environment-Related Communication." *Technical Communication,* vol. 60, no. 2, 2013, pp. 94-117.

Seilhamer, Ryan, et al. "Changing Mobile Learning Practices: A Multiyear Study 2012-2016." *Educause Review,* 2018, https://er.educause.edu/articles/2018/4/changing-mobile-learning-practices-a-multiyear-study-2012-2016.

"Social Media Fact Sheet." *Pew Research Center,* 2019, https://www.pewresearch.org/internet/fact-sheet/social-media/.

"The Role of Technical Documentation." *Astoria Software,* https://2ww.astoria-software.com/the-role-of-social-media-in-technical-documentation.

"Video Viewer Habits, Statistics, and Trends You Need to Know." *TechSmith Corporation,* 2019, https://assets.techsmith.com/Docs/TechSmith-Video-Viewer-Habits-Trends-Stats.pdf.

YouTube for Press, YouTube, https://about.youtube.

Teaching Resources

Overview and Teaching Strategies

The main point of this chapter is that social media platforms can be used as utilities in technical communication, not just for personal entertainment or marketing. As communication mediums they are uniquely powerful for technical writers because they reach such a wide audience of readers or potential readers.

This chapter would be especially useful to teach when students are discussing audience, the types of media used for technical communication, and the purposes for a technical communication project. And while this essay focused on only a few of these platforms when providing examples, students should be encouraged to think about other platforms not mentioned, and perhaps those they use much more frequently.

Above all, the important items to highlight while presenting this chapter are:

- The platform size,
- Its audience characteristics, and
- The type of media it allows users to share.

All this should really challenge students to think about audience and purpose. The three items listed earlier will depend on what the communication is and who it is for. Once students understand that, they will be able to view social media platforms as simply another way to communicate to their readers.

Discussion Questions

To help students explore the ideas discussed in these entries, consider having them address—as individuals, in small groups, or as an overall class—the following questions:

1. As best you can, identify the audience of the social media platform you use most? What type of media is best shared on that platform? Based on these answers, what technical communication purpose would that platform best be suited for and why?
2. Are you able to look at the followers of a certain brand on a social media platform – for example, a car company like Tesla, or a technology company like Apple – and find similarities amongst those people?
3. Is there a form of technical communication, or perhaps a purpose, that social media platforms in general would not be suited for? If so, what type of medium would instead be best and why?
4. Think about what an audience is. Now, discuss how you think social media use in the technical communication industry might change how we define audience. Does it change the relationship between writer and reader? If so, how?
5. Can you identify any new skills or knowledge you might need as a technical writer working with social media platforms that technical writers of twenty years ago might not have needed?

Examining these items can help students better reflect upon and consider how to apply the ideas presented in the essay within the context of their own writing processes.

11 Introduction to Usability and Usability Testing

Felicia Chong and Tammy Rice-Bailey

This chapter describes usability and usability testing.[1] Before we jump into the steps involved in such testing, let's take a look at what we mean by these terms. **Usability** refers to "the quality of a user's experience when interacting with products or systems, including websites, software, devices, or applications. Usability is about effectiveness, efficiency and the overall satisfaction of the user" ("Usability Evaluation Basics"). So, a person's ability to quickly and easily use a texting app to send a text message is a matter of usability—the more easily and quickly they can perform the task, the more usable the app is. For instance, a large, visible button that indicates "Reply," is quicker and easier for someone to use than a menu of options.

Usability testing refers to the task of observing users who are performing planned, purposeful tasks with a product for the purpose of collecting data about how the user interacts with and responds to the product. So, asking a person to use a new app to send a text and then observing how easily and effectively they can perform that task is usability testing. For example, if there is no "Reply" button on the screen, and the user has to scroll through a menu of options to reply, your observing that scrolling behavior (and the possible ensuing frustration) is an example of usability testing.

In usability and usability testing, the term "**users**" refers to individuals who will be interacting with (or *using*) the products (e.g., instructions, websites, mobile apps) we design and develop. Think of usability and usability testing as a research method for gathering information from and about users. The goal is to focus, or center, the design of a product around the individuals who will use it to meet their expectations and needs. A user-centered approach is important in product design and

---

1 This work is licensed under the Creative Commons Attribution-NonCommercial-NoDerivatives 4.0 International License (CC BY-NC-ND 4.0) and is subject to the Writing Spaces Terms of Use. To view a copy of this license, visit http://creativecommons.org/licenses/by-nc-nd/4.0/, email info@creativecommons.org, or send a letter to Creative Commons, PO Box 1866, Mountain View, CA 94042, USA. To view the Writing Spaces Terms of Use, visit http://writingspaces.org/terms-of-use.

development because without it, we are only guessing what will work best for the user. And quite often, these guesses are simply incorrect. The solution is to study the users for whom we are designing products so that our products will meet their needs and expectations. When we design products (whether they be doors, or documents, or mobile applications), it is our responsibility to make sure those products do not unnecessarily complicate the user's life.

To make a product user-centered, you should consider different dimensions (factors) or metrics for measuring usability. Usability expert, Whitney Quesenbery, has identified five dimensions you can use to evaluate the usability of an item:

- Effective
- Efficient
- Engaging
- Error Tolerant
- Easy to learn

To test for usability, you need to answer more specific questions such as those raised by Quesenbery:

- How completely and accurately is the work or experience completed or goals reached?
- How quickly can this work be completed?
- How well does the interface draw the user into the interaction, and how pleasant and satisfying it is to use?
- How well does the product prevent errors and help the user recover from mistakes that do occur?
- How well does the product support both the initial orientation and continued learning throughout the complete lifetime of use?

Thinking about the ways you will measure usability and the types of questions you want to answer will help you more effectively and systematically gather data from representative users.

It is important for technical communicators to understand usability and usability testing. Researchers (Kastman Breuch et al.) have explained that while usability studies and technical communication may use different terminology, the two fields clearly have overlapping emphases, and practitioners in both areas quite often share similar skill sets. Although not all technical communicators will be tasked with conducting usability testing, the principles that this testing employ will help you make decisions that allow you to create positive experiences for the people for whom you create online help, documents, web pages, and other products (your *users*). Also, any document you write for an audience becomes an item that others will use to perform a particular activity or achieve a given objective. For this reason, an understanding of usability and usability testing can help you think about who the readers of your work might be and how you might review and revise your work to meet the needs and expectations of those readers.

This chapter provides you with an introduction to core concepts, tools, and processes associated with usability and usability testing. The objective of the chapter is to help familiarize you with such factors so you can better understand what is involved in these activities. You can then better engage in conversations around these topics and learn more about these processes as you collaborate with others during product development and design activities.

Understanding Users

Definition of User

As we mentioned, users are people who will be interacting with the products you design. Developing an accurate sense of the user is impossible without doing some research. Without this research, you will not have a clear idea of the user's potential inclinations, tendencies, and reactions. Only after you understand your potential users will you have the information necessary to comprehend and respond to questions such as: What confuses them? What frustrates them? What brings them joy? The answers to such questions are key because they allow you to redesign your product accordingly.

Learning About the User

It is not uncommon for us to assume that our users (and perhaps the entire world!) are exactly like us. Manifestly, we know that this is not true, but our instinct is to communicate and design in a way that makes sense to ourselves. However, this egocentric way of communicating and designing sometimes renders products that are confusing or make life more difficult for our intended users. There are several methods by which we can learn about our potential users. Some methods involve working with data about representative users. Other methods require interacting with actual or representative users.

Initial Methods for Understanding Users

Initial methods do not necessarily involve interacting with users; rather, they focus on researching the potential types of users of your product. They are generally used early in the product design process.

User data is gathered to describe the relevant quantitative characteristics (demographic information such as age, gender, location, income, family size) of the type of users you want to target (the individuals for whom you will design something). This data is based on known facts and data about actual users. For example, web analytics (i.e., the collection, reporting, and analysis of website data) may show the main type of user that frequents a website: their age, gender, and location. Alternatively, you might have data about typical users of competitor products.

User data is important because it allows you to identify probable needs and skill levels of particular group your product or document is targeting. When developing products, this data is used at the beginning of the product design process to assist us in creating personas and journey maps, which are covered later. As a result, you can create more usable products because you are not relying solely on your own preferences and desires. Focusing on user data at this beginning stage helps ensure that you are creating a usable product for its intended audience.

Personas are fictional characters based on the real user data that you have collected. They represent a "typical" member of a particular group (e.g., the typical college student from Michigan) and are used to understand how such users would perform an activity or use a product. They answer questions about user motivations, habits, and probable preferences for interacting with the product.

Personas help product developers consider questions such as "What would users from this group do?" when creating products for the members of that group. For example, if you are testing a mobile app to order from a food delivery service, your personas might include a busy stay-at-home mother, a college student living in their dorm, or a wheelchair-bound person who is incapable of driving. You may have questions such as, "What would the typical college student from Michigan do when using an app to order a food delivery to their dorm?"

Creating personas is important because they keep your decisions focused on the intended users of your product or document. Personas give a face to your actual users' potential motivations, fears, concerns, and goals as they relate to the product you are testing. Personas allow you to move past your own assumptions and create products that meet the actual needs and preferences of your user. When developing products, personas are used starting at the beginning of the product design process. Using personas continues to ensure the usability of the product.

Journey maps are graphical representations or visualizations of the process the user follows to accomplish their goals. Essentially, you are charting the path, or sequence of actions, users take as they move through a location in physical space or a series of online screens to perform an activity. For example, a person who is using a mobile app to order from a food delivery service might first select the restaurant from which they want to order food, then select from a menu, and finally complete the transaction with a payment. The journey map for this process would not only establish a timeline of events, but it would also include the potential thoughts and emotions experienced by the user as they go through this process. For instance, this particular user might be excited by the restaurant selection step, overwhelmed by the menu options, and confused by the payment options.

Journey maps are important because they ensure that you will consider and design for the complete range of steps that individuals take while using your product. They also allow you to identify potential pain points (e.g., areas that are difficult to navigate or confusing) for individuals. When developing products, these maps are used throughout the design process. As a result, you can create a more seamless experience for your users.

Advanced Methods for Understanding Users

If we fail to consult or involve representatives of actual users in our product or document design, the decisions we make regarding our design will at best be based on our assumptions of how these users will behave. At worst, our decisions will not even consider the user. Take the classic example (discussed by Norman) of the door that opens by being *pushed*, but has a handle, signifying (erroneously) that it needs to be *pulled*. This example shows how poor design may not only confuse the user but might also make the product appear inoperable.

User data, personas, and journey maps are three tools you can use early in the design process of your product or document. These tools encourage you to keep your users front of mind from the start of your project. They assure that you do not simply project your own needs and desires into the design of your product or document. These tools can also be useful when describing your process to others on the development or management team. Additionally, user data, personas, and journey maps can inform how you select your participants and design your later usability tests. In the following sections, we provide examples of how each of these tools can be implemented and set you up for success on the next steps for designing and conducting a usability test.

Example: Faster Food Mobile User Profile and Personas

Let's say you want to discover how a typical user might interact with the new food delivery application, "Faster Food" that you have developed (see Figure 11.1).

The first task you perform is to create a user profile. For the "Faster Food" app, you develop a user profile that states our users will fit the following criteria:

- Eat fast food
- Have experience using a mobile app
- Be an adult age 18 or over

Figure 11.1: Example of Mobile App

In your user profile, you also identify two categories of users:

1. **Novice users:** Teenagers and adults who are familiar with fast food menus but may not have experience ordering from these restaurants online.
2. **Advanced users:** Teenagers and adults who are both familiar with the fast food restaurants in their area and are experienced using mobile apps to order from such places.

Using this profile and the user categories, you next create two personas, Sean and Roxanne, which you can see in Figure 11.2.

Sean and Roxanne are composites that you created based on data from web analytics that identified demographic information of your average users. Sean and Roxanne's fictional identities include information such as their photos, occupations, yearly income, experience level, and buying habits.

The personas of Sean and Roxanne are important throughout the design and development of your "Faster Food" App to create reliable and realistic representations of your users as you decide what will meet their needs and expectations. For example, in the "Faster Food" app, you might decide to highlight inexpensive restaurants that might be popular with Sean, who has a limited income and lives in the dorms. Alternatively, you might choose to highlight nearby restaurants for Roxanne, who prefers to have her lunch delivered quickly from a restaurant that is near her workplace.

SEAN | 19
Status: Single
Occupation: Sophomore
Yearly Income: $4,000
Experience Level: Novice. Experienced with social media apps but hasn't use an app to order food.
Behavior: Lives in the dorm and never cooks. Sleeps late to finish his homework.
Goals: Prefers fastest service as he spends most of his time getting last-minute meals. Needs affordable, and consistent food.

ROXANNE | 32
Status: Married
Occupation: Administrative Assistant
Yearly Income: $25,000
Experience Level: High. Uses food delivery apps about 2-3 times a week.
Behavior: Often too busy to take a full lunch hour. Usually orders food to be delivered to her office.
Goals: Prefers restaurants that are closest to her workplace for the fastest service during lunch time. Needs highly rated and quality food.

Figure 11.2: Example of Personas

Understanding Testing and Test Plans

Iterative Process of Testing

Ideally, you should start usability testing at the developmental stages of the production process (formative testing), where user feedback is repeatedly gathered and implemented. In other words, the usability testing process should be iterative, where you conduct small studies to learn about user experience, make changes based on what you learned, and then test again. Summative testing is usually conducted later in the product development cycle, when the product is fully developed, to determine whether it meets user requirements. During each stage of usability testing, you want to focus on the user's experience with your product. You want to discover how they initially interact with and react to the product. This will allow you to discern potential questions and pain points for the users. To prepare for each test, you need to develop a test plan.

Purpose of Test Plan

The test plan is a tool that will help you discover how your typical users might interact with your product. A test plan documents your plans for the test, including what you will be testing, who you will be testing, and the tasks you will undertake during the testing.

The purpose for documenting this information is to provide a single source of information that all stakeholders can reference before, during, and after the usability test. Stakeholders are the people who have something invested in the test. This typically includes people designing the test, those administering the test, the project manager, and those financing the test. The final test may deviate from the initial plan, but the changes should also be documented so that the test plan remains up-to-date. As Barnum notes, "you should think of the test plan as a living document that evolves as the materials get fleshed out" (142).

Think-Aloud Protocol

There are many types of usability testing methods depending on the stage of the testing you are in as well as the specific types of data (e.g., qualitative or quantitative) that you are collecting. In this chapter, we introduce a very basic usability test that can be easily performed, the think-aloud protocol, where the participants verbalize their thoughts while they are attempting to use a prototype (a product model that may not include full functionality). This "thinking out loud" allows the tester (the person conducting the test) to understand what the participants are thinking as they engage with the product. It not only allows the tester to identify potential areas of confusion or frustration the participants may be experiencing, but also aspects of the product that are especially pleasing or useful to the participants.

Sections of a Simple Test Plan

Although every test plan is unique, there are several sections that most test plans involve. These include the following:

- Product goal
- Test objectives
- User profile
- Methodology
- List of tasks

Product Goal

The **product goal** identifies the purpose of your product or document. It is the purpose for which the users intend to use your product or document.

Test Objectives

Test objectives articulate the purpose for your test. They answer the question: What do I want to discover from this test? Test objectives might include: How easy/difficult is it for the user to navigate this web page? What are the specific pain points (areas of difficulty) for the user? The test objectives will drive the content and organization of your test.

User Profile

A **user profile** is included in the test plan to effectively articulate who it is that you are looking to test. For instance, this might include demographic information (such as age and gender), job title, level of technical expertise (novice, skilled, or expert), needs and expectations, and steps that are taken to complete a specific task. The earlier work you did analyzing user data and creating personas will be helpful in deciding the types of users that would be appropriate participants in your test. *See the previous Understanding Users section for more details.*

When you do a usability test of a product, you want to select participants (the individuals who will test, or try to use, the product) who match this profile: individuals who meet these criteria and are therefore representative of these users. Only individuals who meet these criteria quality for—or can participate in—this study; those who do not meet these criteria cannot. An example of a user profile might be, "Adults, 18 and over, who use social media and are concerned with privacy." Individuals would then need to meet these criteria to be included in the related study.

Methodology

Your test **methodology** describes how you will gather data, more specifically, the number of participants that you need, length of the test, and test procedure. To ensure that you gather the type of data that will be most useful to you, it is important that you provide your participants with specific tasks to perform.

List of Tasks

These **tasks** are often embedded within scenarios that provide the context for the user's goals. For instance, if you are interested in how a user might interact with your e-commerce site, you could develop a scenario in which the participant has to find and place orders in an online shopping cart. One specific task involved in this scenario might be to find a particular item, say, a blanket. As Barnum notes, "without a common set of scenarios, users will go their own way in an interface, which makes it difficult to see patterns of usage and recurrence of problems among and between users" (19).

Example: Faster Food Usability Test Plan (Simple)

The first part of the test plan (Figure 11.3) is to articulate the product goal and test objectives. As you can see, your objectives are to verify the navigation within the app is clear and to determine how enjoyable the app is to use. Next, you will create a user profile based on the data that you have previously obtained about actual users. Your methodology is a simple, think-aloud protocol. Next, you will document the list of tasks the participants will complete, which includes:

1. Navigate from the homepage to the restaurant selection page.
2. Return to the home screen from a restaurant page.
3. Navigate to the pay screen.
4. Navigate to the home address screen.

> **Faster Food Mobile App Test Plan (Simple Version)**
>
> **Product Goal**
> Help users order food from local fast food restaurants
>
> **Test Objectives**
> - Verify the navigation within the app is clear.
> - Determine how enjoyable the app is to use.
>
> **User Profile**
> - Eat fast food
> - Have experience using a mobile app
> - Be an adult age 18 or over
>
> **Methodology**
> Using the think-aloud protocol, we plan to test a total of five participants (two novice and three advanced), who will verbally describe their experience interacting with the Faster Food app. Their descriptions will be recorded by a note-taker. Each test will be 15 minutes.
>
> **List of Tasks**
> 1. Navigate from the homepage to the restaurant selection page.
> 2. Return to the home screen from a restaurant page.
> 3. Navigate to the pay screen.
> 4. Navigate to the home address screen.

Figure 11.3: Example of Simple Usability Test Plan

Additional Sections Included in an Advanced Test Plan

An advanced test plan would contain more comprehensive tasks and test details, including different types of qualitative and quantitative data (**evaluation method**) you are seeking and specifics about the testing environment (**description of testing environment**), as described later in this chapter.

Evaluation Method

There are multiple ways to collect both quantitative and qualitative data, depending on whether you are conducting formative or summative testing. There are two types of quantitative data: **Performance data** is based on measurements of users' actions, whereas **preference data** is based on user's responses to questions on questionnaires. Qualitative data includes observations of user actions and user comments (during the test or on questionnaires). You can collect either type of data using the following methods.

Audio and/or video recording. For example, if you are testing a website, you can use a screen-recording program to record the user's on-screen activities (e.g., where they click or how long it takes them to navigate).

Verbal feedback protocol. There are various ways to solicit qualitative feedback from your users using the think-aloud protocol (e.g., concurrent think aloud, retrospective think aloud, concurrent probing, or retrospective probing, as described in "Running a Usability Test"). You can also use product reaction cards, which were developed by Microsoft (Benedek and Miner) to provide a rich understanding of user experience.

Questionnaires. The three common types of questionnaires include pre-test (e.g., on user preferences), post-task (e.g., on issues such as ease or difficulty, after each task), and post-test (e.g., on the overall experience at the end of the test).

Barnum recommends combining "metrics with your observations, comments, from participants, and their responses to open-ended questions" (137) to create a highly effective test.

Description of Testing Environment

The description of your testing environment details the physical conditions under which the testing will occur. For instance, you might be observing an individual using a new point of sale (POS) app in a fast-food restaurant. Alternatively, you might be observing an individual using the same app in a testing lab. Each of these environments is going to be different. The noise level, lighting, level of activity, and other considerations might have an impact on your test.

Example: Faster Food Usability Test Plan (Advanced)

In an advanced test plan of the Faster Food app, you will include the additional sections such as evaluation method and description of testing environment. See Figure 11.4 and Figure 11.5 for an example of an advanced test plan.

Pilot Test

A pilot test is a small preliminary study used to test a proposed research study before conducting a full-scale study. This preliminary study typically follows the same procedures as its full-scale counterpart. The primary purpose of a pilot study is to evaluate the feasibility of the proposed study. A pilot test is sometimes called a field test or a feasibility study.

A pilot test will allow you to see if you are testing for the correct questions. Pilot tests also allow you to be sure you have considered logistical implications and established best practices for the full-scale usability test. For example, researchers (Zimmerman et al.) who were investigating rural health and safety issues planned a pilot in which they tested pesticide warning labels at a university's horticultural research farm. Although they believed they had sufficiently prepared for the usability tests, the researchers encountered a few logistical problems including a thunderstorm. They then knew they had to revise their full-scale study and equip the testing team with adequate clothing to include a better hat and protective boots.

Recruiting Participants

Often, students wonder how many participants they need for a usability study. Nielsen makes a sound argument for aiming for three to five users (Nielsen). They explain that you will learn the most from the first user. The second and third user will do many of the same things that you already observed with the first user. The third user will generate a small amount of new data, but not as much as the first and second user. As you add users, you will continue to observe the same behaviors and will learn less with each additional user. As Nielsen explains, "After the fifth user, you are wasting your time by observing the same findings repeatedly but not learning much new."

Faster Food Mobile App Test Plan (Advanced Version)

Product Goal
Help users order food from local fast food restaurants

Test Goals
- Test that the buttons and icons on the app are consistent and intuitively positioned.
- Test that the text is readable and understandable to the users.
- Test that the costs/fees and delivery times are clearly visible to the users.
- Verify the navigation within the app is clear.
- Determine how enjoyable the app is to use.

User Profile
- Eat fast food
- Have experience using a mobile app
- Be an adult age 18 or over

Methodology and List of Tasks
Number of participants:
We plan to test a total of five participants: two are novice and three are advanced users.

Length of the test:
The total length of each session will be 30 minutes, including:
- Introduction: 5 minutes
- Task scenarios: 20 minutes
- Debriefing: 5 minutes

Figure 11.4: Advanced Test Plan Part 1

> **Test procedure:**
> The test will begin with an introduction of the test, followed by scenarios described below, and end with debriefing.
>
> Scenario 1: You want to order a Big Mac value meal from McDonald's, and you want to select a large Diet Coke as your drink. Please show us how you would select the closest McDonald's to you with the highest ratings and place the items in your cart to checkout. Then, edit your cart by changing the Diet Coke to a Sprite. Provide a special delivery instruction to come to the East entrance of the delivery address. Go back to your cart and checkout by entering your address and credit card information.
>
> Scenario 2: You want to order a Greek salad and a large Passion Papaya Green Tea from Panera Bread. Please show us how you would select the closest Panera Bread to you and place the items in your cart to checkout. Then, go to the dessert menu and add a chocolate chip cookie. Checkout by entering the delivery address and your credit card information.
>
> **Testing environment:**
> The test for the Faster Food mobile app will be conducted in a classroom on our campus, using the mobile phone provided by the researcher. Recording of sessions will be done using the webcam on a laptop. The moderator and notetaker will be seated nearby.
>
> **Evaluation Method**
> **Qualitative data:**
> Participants will follow the think-aloud protocol, where they verbally describe their experience using the app during the test. Their descriptions will be recorded, and the notetaker will make a note of the comments. After each scenario, users will describe what they enjoyed the most or the least about the app and provide recommendations for improving the app (e.g., on visuals, text, information such as costs and delivery times, and navigation).
>
> **Quantitative data:**
> In addition to the qualitative data, we will also look at the specific number of times certain events occurred, including the following:
> - Successful task/scenario completion rate
> - Number of critical errors users make that prevent them from finishing their tasks
> - Number of non-critical errors users make that affect their efficiency (e.g., using the "back" buttons)
> - Amount of time it takes the users to complete the task
> - Ease of use and satisfaction of the app on a 5-point Likert Scale

Figure 11.5: Advanced Test Plan Part 2

One of the greatest challenges to conducting a usability test is recruiting participants. Research (Chong, "Implementing Usability Testing") shows that nearly one quarter of students reported that finding users to participate in a usability test was a challenge. To make this process more manageable, it is important that you define specific criteria for potential participants and think broadly about the recruitment process.

Establish Criteria

Before you start recruiting participants for your usability test, you want to be sure that you have established criteria that potential participants need to meet. You will then use these criteria to screen potential participants. For instance, if you are testing user instructions for a riding lawn mower, you probably want participants who know how to both mow a lawn and drive some type of vehicle. Selecting a participant who is missing either of these qualifying criteria will not provide you with valid or representative results about how an actual, typical user of your instructions will react. Once you have established criteria that participants need to meet, you are ready to recruit.

Identify Sources for Participants

There are several sources from which you can recruit participants for your test. One way is to use your personal and professional networks. As Barnum explains, making requests "through friends, family, business associates, and community connections often leads to qualified applicants for screening" (159). Very often, if a connection you make through one of these channels meets your screening criteria, it is possible they have people within their networks that will also meet your criteria. In this sense, you can often find potential participants via other participants. If you have a limited network, or if you do not have many in your network who will likely meet your criteria, you can research organizations that likely have as members people who fit your desired demographic. For instance, if you are looking for people to test lawn mower instructions, but most people in your network live in apartments and don't have the need to mow a lawn, you might solicit local homeowners' associations for names and contact information of potential participants. If you are conducting a usability test for an organization, and they have the budget for external resources, you may decide to hire a market research firm to assist you with your recruitment process. Such firms generally have substantial databases of potential participants who can be organized by particular demographics and can be further interviewed (by the firm) for more specific criteria.

Screen Potential Participants

Just because an individual has met your initial screening criteria, does not mean that person is right for your usability test. Before scheduling someone to participate in your test, you will want to confirm (either by phone or e-mail) that they have, indeed, met your initial criteria. You might also follow-up with additional questions that could disqualify the person. For instance, you probably do not want to select someone who builds lawn mowers—and so would use their own knowledge (instead of your instructions alone)—to evaluate your instructions. You will also want to confirm that your potential participant is available during the time you plan to conduct the usability tests.

Preparing Test Materials and Environment

Once you have designed and piloted your test, along with recruiting the appropriate number of participants, then you need to prepare test personnel, materials, and the environment for your test. While a simple test using the think-aloud protocol may not have formal roles and responsibilities for the testers, it is still important that you familiarize yourself with the full range of testing roles and tasks.

Roles and Responsibilities

If you have a team working on the usability test, there should be at least be two roles:

1. Moderator/facilitator: This person will interact the most with the participants by asking unbiased questions and guiding them through the test. For detailed information about the role of the moderator/facilitator, see the *Conducting the Usability Test* section (later in this chapter).
2. Observer/notetaker: If your test is not recorded (e.g., audio or visual), you will need a note-taker who will serve as an observer and recorder of the test.

Team Checklists

Based on the roles that you have assigned for the test, you will need to create checklists that help remind the team members of their responsibilities and ensure that each test session is consistently conducted. For example, the moderator's checklist may include tasks for welcoming the participants, reviewing the consent forms, and providing instructions for the test. The notetaker's checklist may include tasks that need to be completed before, during, and after the test. Loranger provides a nine-step checklist for planning a usability test.

Moderator's Script

The goal of having a moderator's script is to ensure that the moderator explains the process in the same or consistent manner to each participant. The script will cover topics such as welcoming the participant, stating the purpose of the study, providing the consent form, explaining the testing process, and describing the think-aloud protocol. It is especially useful to use phrases such as "we are testing the product, not you" before the test to make the participants feel comfortable and less self-conscious about describing their challenges.

OTHER MATERIALS

Depending on your test plan, you may include other materials such as the following:

- Consent forms that indicate the participant's willingness to participate or be recorded during the test;
- Non-disclosure agreements, if you are testing a product in development;
- Observer/note-taker form on which to record the issues/problems that each participant experiences on each scenario/task;
- Questionnaires to obtain qualitative and quantitative feedback from participants that will enhance your understanding of the user experience;
- 118 product reaction cards created by Microsoft (Benedek and Miner), which include adjectives such as "confusing," "difficult," "engaging," and "entertaining" that the participants select to describe their experience with the product being tested.

ENVIRONMENT

Based on the description of the test environment that you have included in your test plan, you will need to make sure that that space (whether physical or online) is ready for testing. For instance, if the testing will be done in a lab or classroom, you need to be sure the camera is functioning and recording, that there is sufficient seating for participants and observers, and that the noise level is not prohibitive. Similarly, if you are conducting an online test, then you need to be sure that internet connectivity, video cameras, and audio are functioning properly.

CONDUCTING THE USABILITY TEST

The moderator's role is critical in that they will need to conduct the usability test effectively. In our experience, this is the most challenging and critical role because the moderator needs to create a comfortable environment and listen to the participant's feedback carefully (e.g., to ask follow-up questions). While a simple test using the think-aloud protocol may not differentiate between tasks for the moderator and notetaker, it is still important that you familiarize yourself with the full range of these tasks.

Moderator Tasks

Immediately prior to and during the usability test, the moderator has several tasks, which may include the following:

- Meeting, greeting, and briefing the participants.
- Preparing the participants for the feedback method or protocol (e.g., using concurrent think aloud, retrospective think aloud, concurrent probing, or retrospective probing, as described in "Running a Usability Test").
- Being effective and unbiased by monitoring both your verbal and nonverbal communication and asking "what" or "how" (and not "why") questions to get the participants to share their thoughts with you.
- Knowing how and when to intervene (e.g., when the technology crashes, when the participant struggles to answer the question, or when the participant wanders off tasks).
- Distributing test materials (e.g., questionnaires, product reaction cards).

Considerations for Moderators

The first few times you act as a moderator, you may find yourself experiencing the "stupid user syndrome" (see Chong, "The Pedagogy of Usability"; Johnson). Because you are familiar with (and may have even helped develop) the product, you believe that it is intuitive and easy to use. When a user struggles with the product, your first inclination may be to assume it is the fault of the user. It is important to be aware of the appearance of the "stupid user" syndrome, where the moderator may blame the participant for struggling with their tasks or wandering off of the tasks.

Likewise, moderators should avoid leading, confrontational, or unbalanced questions because these types of questions will not give you an accurate picture of your user's experience with the product. For instance, if you ask, "Wouldn't you agree that this button is misplaced?" you are suggesting that you already know what the user is thinking (and you may be wrong!). You could instead, revise this question to, "Where do you expect to see this button?" Examples of other questions you should *avoid* include the following:

- How difficult was that task for you? (Revised: "How easy or difficult was that task for you?")

- Do you like the instructions? (Revised: "How did the instructions help you reach your goal?")
- Why did you click that button? (Revised: "What made you click that button?")

Considerations for Notetakers

The notetaker's job is to log observations of the participant interacting with the product. Typically, the notetaker does not interact directly with the participant, but occasionally, the notetaker may ask clarifying questions (so that they can accurately record the participant's reaction or statement). In this sense, the notetaker is largely an observer. Some usability tests use multiple notetakers. This enables the testing team to correlate findings after the test. However, some tests also include other individuals whose sole job is to observe. For example, if you are conducting a usability test for a corporation, members of the management team or product development team may simply observe the testing (but not take notes or perform any other specific task).

Analyzing the Findings

After conducting the test, there are three main questions that need to ask when analyzing the data (Barnum 239):

- What did you see?
- What does it mean?
- What should you do about it?

Similar to analyzing the primary research data in other projects, you will need to begin by gathering the data from everyone (e.g., moderator, notetakers, and observers). After gathering all your data, you will analyze your findings. Having more analysts will capture more problems and yield better findings (Sauro).

Example: Analyzing the Findings of the Faster Food Mobile App (Simple)

To analyze the results of the "Faster Food" app simple usability test, you look at the data you collected from the participants' think-aloud protocol statements. This data is found in the notes or transcript of the participants' statements. Figure 11.6 shows these comments from three of the five participants who followed the think-aloud protocol while they completed our usability test.

Participant	Screen	Comment
Roxanne	home	I like the color scheme for these icons.
	restaurant	How do I find out the ratings for this McDonald's? Oh, here it is.
	menu	How do I change my drink order to a Sprite?
	order	It is easy to put in the special instructions.
		Do they take Apple Pay? I guess not.
Sean	restaurant	The list is sorted by distance, so I always start with the closest McDonald's to campus. Nice.
	order	A lot of scrolling to find the drinks.
		I can save my credit card information, which is nice.
Trisha	home	Nice images.
	restaurant	I don't see McDonald's. Wait. next page.
	menu	Wow! It looks like the full menu is available for delivery.

Figure 11.6: Example of Qualitative Findings

By creating a column that identifies the screen the participant was on when they made the comment, you could look for patterns in the data. For instance, Figure 11.6 shows that while on the restaurant selection screen, two of our participants initially questioned how to find a specific restaurant. This also shows that two participants gave favorable feedback on the home screen.

Example: Analyzing the Findings of the Faster Food Mobile App (Advanced)

For a more advanced usability test, which might include quantitative data, you would review this quantitative data in addition to the qualitative data described in the simple usability test. For example, you look at the specific number of times certain events occurred during the testing. Figure 11.7 indicates the events you are testing and the participants' scores:

Figure 11.7 shows some of the quantitative data from the usability tests. Notice that all three participants spent the least amount of time selecting the restaurant and the most amount of time changing the Diet Coke to a Sprite. Also, notice that the task of selecting the Big Mac Meal shows the most occurrences of using the "back" button, which may suggest that this screen is not as intuitive or user-centered as you would have expected.

Task	Number of times using the "back" button	Number of times returning to home page	Length of time to complete	Ease of use (1-5; 1=very unsatisfied, 5=very satisfied)
Task 1: selecting the highest rated McDonald's in the area				
Roxanne	0	0	20 sec	4
Sean	1	0	22 sec	5
Trisha	1	0	31 sec	4
Task 2: selecting the Big Mac value meal				
Roxanne	1	0	40 sec	3
Sean	1	0	30 sec	4
Trisha	2	1	110 sec	2
Task 3: changing the Diet Coke to a Sprite				
Roxanne	0	0	42 sec	4
Sean	1	1	128 sec	1
Trisha	0	0	117 sec	3

Figure 11.7: Example of Quantitative Data

Developing Recommendations

Once you gather and analyze the data from the usability testing, you will develop recommendations based on this data. For instance, if your data indicates that participants spent more time on a particular screen of your mobile app than you anticipated, you may recommend that the screen be redesigned. More specifically, if the data suggests a specific issue with that screen (the buttons were difficult to locate), you might recommend increasing the size of the buttons or adding a brighter color.

Your recommendations may be informally or formally presented depending on your audience (the person or group for whom you are conducting the usability test). If you are conducting a simple, think-aloud usability test, you may summarize your findings and recommendations in a couple of short paragraphs. If you are conducting an advanced usability test, you may be asked to create an executive summary of your recommendations. The executive summary simply documents the issue and your recommended fixes. The test data and your analysis, which support your recommendations, can be attached to the executive summary so that they are available, if needed.

Example: Developing the Executive Summary for the Faster Food Mobile App Usability Testing

Figure 11.8 shows the executive summary of the "Faster Food" app advanced usability testing. This summary includes an introduction to the project, a brief summary of our methods and findings, and the recommendations that these findings suggest.

Executive Summary: Faster Food Mobile App Usability Testing

Introduction
The Faster Food mobile application assists users who want to order food from local fast food restaurants to be delivered to them. Our users are adults aged 18 and older who have experience using mobile apps, though this experience level may vary. Our usability testing involved five representative users between the ages of 19 and 50. The testing was performed in our Rochester, Michigan lab.

Summary of Methods
We used a combination of qualitative and quantitative methods to gather our data. Qualitative participant data was gathered by recording the statements made while the participants were following the think-aloud protocol while completing Scenario 1 of our test. Quantitative participant data was gathered by looking at the number of times certain events happened during the testing, such as the participant selecting the back button or returning to the home page.

Summary of Findings
Positive Findings
- Four of five users found the app attractive, two citing the "color scheme" and "images."
- All five users found the checkout process easy, where they can easily enter their addresses, special instructions, and credit card information.
- Three of the five users found that selecting the desired restaurant was straight-forward.

Negative Findings
- Four of the five users found that the ordering screen was difficult to navigate, and there was a higher-than expected use of the "back" button on this screen.
- Three of the five users found making changes to the orders to be cumbersome.

Recommendations
- Due to the high number of participants commenting on the attractiveness of the app, we recommend retaining the color scheme and icons currently being used.
- To address the number of times the participants use the "back" button on the ordering screen, we recommend adding a large button that reads "change my order."
- Because some of the users had a difficult time finding the closest, highest-rated restaurant to them, we recommend adding an algorithm that default to this selection on the restaurant selection screen.

Figure 11.8: Example of Executive Summary

Conclusion

In this chapter, we explain the importance of understanding users and provide basic guidelines for designing and piloting a test plan, recruiting participants, preparing test materials and environment, conducting the test, analyzing the findings, and developing recommendations. We provide examples of both a simple, think-aloud protocol usability test (that includes qualitative data) and a more extensive usability test (that includes both qualitative and quantitative data). In summary, usability testing is a crucial tool to add to your skillset as a technical communicator, and it is important to solicit as much user input and as often as possible during the product development process.

Works Cited

Barnum, Carol. *Usability Testing Essentials: Ready, Set . . . Test!* Elsevier, 2011.

Benedek, Joey, and Trish Miner. "Measuring desirability: New methods for evaluating desirability in a usability lab setting." Proceedings of UPA Conference, Orlando FL, 2002.

Chong, Felicia. "Implementing Usability Testing in Introductory Technical Communication Service Courses: Results and Lessons from a Local Study." *IEEE Transactions on Professional Communication*, vol. 61, no. 2, June 2018, pp. 196–205. https://doi.org/10.1109/TPC.2017.2771698.

Chong, Felicia. "The Pedagogy of Usability: An Analysis of Technical Communication Textbooks, Anthologies, and Course Syllabi and Descriptions." *Technical Communication Quarterly*, vol. 25, no. 1, 2015, pp. 12-28. http://doi.org/10.1080/10572252.2016.1113073.

Johnson, Robert R. *User-Centered Technology: A Rhetorical Theory for Computers and Other Mundane Artifacts*. State University of New York Press, 1998.

Kastman Breuch, Lee-Ann, Mark Zachary, and Clay Spinuzzi. "Usability instruction in technical communication programs: New directions in curriculum development." *Journal of Business and Technical Communication*, vol. 12, no. 2, 2001, pp. 223-240.

Loranger, Hoa. "Checklist for Planning Usability Studies." *Nielsen Norman Group*, 17 Apr. 2016, http://www.nngroup.com/articles/usability-test-checklist/.

Nielsen, Jakob. "Why You Only Need to Test with 5 Users." *Nielsen Norman Group*, 18 Mar. 2020, http://www.nngroup.com/articles/why-you-only-need-to-test-with-5-users/.

Norman, Donald A. *The Psychology of Everyday Things*. Basic Books, 1988.

Quesenberry, Whitney. "Using the 5Es to Understand Users." WQusability, http://www.wqusability.com/articles/getting-started.html.

"Running a Usability Test." Usability.gov, http://www.usability.gov/how-to-and-tools/methods/running-usability-tests.html.

Sauro, Jeff. "How Large is the Evaluator Effect in Usability Testing?" *Measuring U*, 27 Mar. 2018, http://www.measuringu.com/evaluator-effect/.

"Usability Evaluation Basics." *Usability.gov*, http://www.usability.gov/what-and-why/usability-evaluation.html.

Zimmerman, Donald E., Michel L. Muraski, and Michael D. Slater. "Taking Usability Testing to the Field." *Technical Communication*, vol. 46, no. 4, 1999, pp. 495-500.

Teacher Resources

Overview and Teaching Strategies

There are various models for implementing a usability testing project in the classroom. Ideally, this project should be taught in a dedicated usability course to allow students time to develop the skills involved in usability testing. If, however, you must implement the usability test as one of the major projects as part of another class (such as technical writing), we recommend that you identify one project or a reliable or responsive client for all student teams (instead of trying to manage several projects/clients).

Just as with other types of primary research, students must learn how to effectively collect user data and analyze it, so it is crucial for you to support them by giving them timely and useful feedback (e.g., on their test plans) prior to the test. Also, be sure to allow ample time for all phases of the project, as many may underestimate the amount of time required for various tasks (such as recruiting and scheduling participants) and the iterative nature of other tasks (such as analyzing data). It is also important to select participants for the usability test who match the user profile, but it is often a step that is easily skipped or underestimated. Therefore, you should allow ample time for students to recruit users.

Discussion Questions

Following are discussion questions that will allow your students to explore the ideas discussed in this chapter:

- What kind of primary research studies have you participated in in the past?
- What type of experience do you have creating personas?
- What do you think are the major challenges involved in conducting a usability study?
- What methods are you most likely going to use to recruit participants?
- What are some current issues on your campus or in your field that could be investigated with this usability testing method?
- How could usability fit into your major field of study?

Suggested Websites for Remote Testing

With the advent of online education and testing, here is a list of websites for conducting tests online or remotely:

- Usabilityhub.com

- OptimalWorkshop.com
- UserTesting.com

Suggested Readings

Here is a list of suggested readings for you to learn more about user-centered design and usability tests:

Barnum, Carol. *Usability Testing Essentials: Ready, Set . . . Test!* Elsevier, 2011.
Johnson, Robert R. *User-Centered Technology: A Rhetorical Theory for Computers and Other Mundane Artifacts.* State University of New York Press, 1998.
Norman, Donald A. *The Psychology of Everyday Things.* Basic Books, 1988.
Rubin, Jeffrey, and Dana Chisnell. *Handbook of Usability Testing: How to Plan, Design, and Conduct Effective Tests*, 2nd ed., Wiley, 2008.

12 Beyond Audience Analysis: Three Stages of User Experience Research for Technical Writers

Joanna Schreiber

Writing for your audience is foundational to all technical writing work.[1] Understanding the audience is particularly important for technical writers who must also advocate for audience's (users') needs. You have probably been asked to analyze the intended audience before, especially when crafting a message or writing a paper. But what do you really know about the intended audience and how do you know it? A multi-phase and systematic approach to user research, as described in this chapter, can help you understand audiences and develop effective products and services for them.

Advocating for audiences requires "considering our users' ways of making meaning and privileging our users' experiences and worldviews by inviting our users into the design process" (Jones 317). Social justice approaches to technical writing and engineering illustrate ways in which design can disenfranchise users (e.g., see Dorpenyo; Gonzales; "Building Access"; Leydens and Lucena; Jones and Williams; Sanchez). Technical writers must position users' experiences, physical and cognitive abilities, and cultural backgrounds as integral to effective design practices.

As advocates for users, technical writers need to know how to effectively learn about and work with audiences, but many audience analysis exercises rely on assumption rather than research (Lam and Hannah). User experience (UX) research practices provides a range of methods and tools for learning about audiences (groups of users) and how they engage with products and services. As Hackos and Redish explain, going beyond

---

1 This work is licensed under the Creative Commons Attribution-NonCommercial-NoDerivatives 4.0 International License (CC BY-NC-ND 4.0) and is subject to the Writing Spaces Terms of Use. To view a copy of this license, visit http://creativecommons.org/licenses/by-nc-nd/4.0/, email info@creativecommons.org, or send a letter to Creative Commons, PO Box 1866, Mountain View, CA 94042, USA. To view the Writing Spaces Terms of Use, visit http://writingspaces.org/terms-of-use.

assumptions about users to actually learning about users is an important skillset for technical writers:

> We need to study users because users decide whether to use a product, not designers or writers. Even if the users' supervisors can dictate what they must use, the way people use products is self-determined. We also need to study users because the more we know about them, the better we can design for them. They are people with likes and dislikes, habits and skills, education and training that they bring into play whenever they work with a product. (Hackos and Redish 25)

Users are people who bring their own preferences and needs to each product, service, and communication situation. What Hackos and Redish point out is that technical writers do not know enough about the needs, values, and goals of users from their own experiences and may inadvertently incorporate their own biases by relying on their personal assumptions. Conducting UX research helps technical writers, and other members of design teams, to move from relying on their individual assumptions and toward working with real audience perspectives (Johnson; Smith et al.). Bringing these perspectives into the design process is incredibly important for effective and inclusive design. As Acharya explains, "building an inclusive and just future starts with understanding the needs and expectations of all users, including underserved, underrepresented user groups from different cultures" (17).

User experience (UX) research is an extension of audience analysis that incorporates a range of research methods, seeks to include user perspectives, and positions users as experts. As part of the process for developing content and designing products for users, UX research situates a range of audience concerns, including audience goals, accessibility, and context of use. Technical writers—who design information to meet audience needs and goals—must think from a UX perspective in order to design content and navigation to support users. Without building knowledge from a user perspective, it is easy to misinterpret user needs or exclude them altogether.

Don Norman illustrates the importance of resisting assumptions about audiences and learning about audience from the perspective of a user feeling excluded product designs. He wrote about his experience as an "elderly user," describing products designed for his age group as ugly and difficult to use:

> Despite our increasing numbers the world seems to be designed against the elderly. Everyday household goods require knives and

pliers to open. Containers with screw tops require more strength than my wife or I can muster. (We solve this by using a plumber's wrench to turn the caps.) Companies insist on printing critical instructions in tiny fonts with very low contrast. Labels cannot be read without flashlights and magnifying lenses. And when companies do design things specifically for the elderly, they tend to be ugly devices that shout out to the world "I'm old and can't function!" We can do better. (Norman)

Norman is describing products as misunderstanding the needs and values of his age group, in terms of accessibility, functionality, and aesthetic. The range of issues Norman describes affects product design, user support, user orientation, interface design, information design, and typography. Norman's essay illustrates that design decisions targeting his user group are based on assumptions and not research. Learning about users and listening to users are necessary first steps to avoiding assumptions and being able to advocate for users.

This chapter presents UX research as a three-stage process to help you build on what you already know about audience analysis and to help you effectively incorporate data collection, research methods, and user documentation into technical writing processes. The three stages of user research presented here are background research, primary research, and usability testing. Background research involves collecting and analyzing existing studies (published research), available market research, datasets, and artifacts. Primary research involves designing a study to collect data that has not been collected in previous studies or to update findings from previous studies. Usability testing involves testing a specific element of a design (content, feature, navigation) with your audience.

These stages are iterative and flexible. That means that as you progress through the stages, you may continue to conduct background research (Stage One) and perhaps incorporate user tests (Stage Three) as needed throughout the design process. The first two stages help you narrow what to design and test in the third stage. For example, the first two stages might help you design the progression and instructions in an app or the navigation of a website, and the third stage can help you test those features in order to refine them. At the end of each stage, you'll organize your findings and make recommendations for next steps. This chapter does not cover each deliverable you can develop at each stage—including personas, scenarios, and journey maps—or explain each method you can use at each stage. It gives technical writers a framework for conducting UX research with content development and information design work.

From Audience Analysis to UX Research

Audience analysis is too often reliant on the writer's experience and intuition (Lam and Hannah 29). Even when audience analysis begins with identifying demographic characteristics like age, marital status, income, etc., relying solely on these characteristics often positions the technical writer to make too many assumptions about their audience based on limited information. For example, date of birth or income often tells you little about technology skills, cultural background, preferences about adopting technology (e.g., early adopters), or what kinds of features will make a product more accessible. Learning more about these audience values and goals will help a technical writer craft better technical content, provide better task orientation features, and design more effective navigation and visual information.

To emphasize the limitations of demographic information, consider the once-popular meme about Ozzy Osborne and then Prince Charles. They are very different people but have many demographic characteristics in common. They are both the same age, both from the same country, both have remarried, both have children, and both are wealthy (Ward). But, as marketing analysts have pointed out, this information is helpful, but does not offer a complete in understanding how best to address them as users. In other words, while these two users may belong to the same age group, they are not likely to belong to the same user group.

This example shows the complexity of audiences that cannot be explained solely through demographic characteristics like age, income, and nationality. Technical writers must be able to go beyond these characteristics to find out more about the values, goals, interests, and cultural backgrounds of users. Through UX research, technical writers can systematically build information and assess their research practices in order to properly support, work with, and advocate for users.

Advocating for users is an important technical writing responsibility, particularly to ensure products (including content) are inclusive and accessible (Jones; Rose and Schreiber). Accessibility should not be seen as compliance, but as a shared value ("Building Access"; "Universal Design and the Problem of 'Post Disability' Ideology"; Huntsman). Accessible design supports a spectrum of cognitive and physical abilities. As the World Wide Web Consortium (W3C) points out, web accessibility supports a range of users including older users, users with disabilities, and users in rural areas. Huntsman explains accessibility from a technical writing perspective:

> Every designed communication is a built environment with methods of access. The words we choose, the format we design, and the technology we use are all elements that develop access to information. (Huntsman 233)

Huntsman's definition positions design as a series of choices, and all of those choices affect accessibility. It is problematic when design teams, including technical writers, make assumptions about users and choices about designing for users from those assumptions. Instead, technical writers need to infuse their work with UX research practices so that at all stages of writing and design, they can mitigate inaccessible design choices by learning about and with users.

As Huntsman shows, accessible design advocacy is about both engaging with design decisions and with audience needs and values. Advocating for users and promoting inclusive design begins with listening to users (Smith et al), valuing their experiences and cultural backgrounds (Jones), and identifying choices in design and opportunities for intervention and inclusion.

Three Stages of User Research

Technical writers use UX research practices to collect information about audiences and test content, service, and product features. In other classes, you may have learned about primary research (collecting new data) and secondary research (consulting existing studies). Because secondary research tells you what has already been studied relating to your research questions, it helps you decide what new data needs to be collected or what data needs to be updated (primary research). These three phases—background research, primary research, and usability testing—show how secondary research sets up primary research by helping the researcher systematically identify underexplored areas of research or studies that need to be revisited. The primary research stage sets up the more usability testing stage, which is a more focused type of primary research.

Technical writers often work in collaborative environments, often work on multiple projects, and often work on both short-term and long-term projects. Therefore, it is incredibly important for technical writers to document their research findings and make recommendations at each stage. This will help others understand how they collected their data, help others on the project use the data, help managers and project managers make decisions using the data, and help others collaborate with the data collection. That is why each stage describes the research to be done as well as how to organize the findings and write recommendations. The findings of research at each stage need to be organized in a way that can be used by others as well as help the technical writer track what they've learned.

As you move through these stages, it's important to regularly pause and consider what perspectives and experiences are being left out and how to best directly include those perspectives. Because technology often has embedded racial, gender, and ability bias (Broussard), regularly trying to incorporate additional perspectives is important. In their argument for direct collaboration among designers and people with disabilities, Oswal and Palmer argue that including users with disabilities:

> . . . employ their distinct know-how about disability which originates from their bodily differences and diverse contexts of purpose and use. In the case of "context of use," design work with disabled users differs significantly from design work with other users because most participatory studies do not focus on this aspect of design. Disabled people bring viewpoints of their own of being in and with the material and social world which shape, at least in part, their human desires, needs, and expectations. Disabled bodies traverse through these worlds at a different pace, in diverse ways, and for succinct purposes to fulfill these needs, desires, and worldly goals which might appear odd, out of place, or even undesirable to a non-disabled eye and a presumably fit body. (Oswal and Palmer 251)

The immersed disability perspective Oswal and Palmer describe helps technical writers and designers understand both user goals and context of use (Stage Two).

These three stages are designed to help technical writers systematically gather information about users, to involve users directly in design, to listen to users, and to treat users with respect and dignity. While every approach will have flaws, focusing on building reciprocal relationships with user groups, particularly marginalized communities, helps minimize issues (Cardinal et al; Smith et al).

Research Stage One: Background

The first stage of UX research is to gather information from previous studies or existing artifacts: background research. Conducting background research is a focused activity, and technical writers should start with studies that are specific to a group of users and studies specific to the technology or service being developed, including accessibility concerns.

To begin, technical writers need to identify at least one audience or group of users. You may be studying multiple audiences and should gather information about each separately before synthesizing their findings and recommendations. As technical writers, you need to decide what you want to know about the audience (research questions). The research question(s) becomes important because it helps you identify key words and focus your search. Your research question helps you define your group of users as well as related technology, product or service questions. It's okay if your research question is general, but you want to make sure you push beyond demographic questions to get at values and needs of users.

For example, a group of users might be adults 65 and older. This is the user group that Donald Norman discusses as being underserved at the beginning of this chapter. It is a very larger user group, spanning multiple generations, and as you conduct background research, you may wish to divide this large user group in to multiple user groups based on what you learn. Possible research questions to get you started conducting background research about adults aged 65 and older and their technology use include:

- What do adults aged 65 and older value about technology?
- Why do adults aged 65 and older adopt technology?
- What technologies do adults aged 65 and older adopt?
- What accessibility or design features have been developed to serve adults 65 and older?
- What accessibility or design features have been tested with adults aged 65 and older?

These example research questions address accessibility, technology or service adoption, and value separately. By asking them separately, you will be able to find studies related to each of these items, which may or may not overlap and identify a wider range of key phrases. For example, finding studies about what the audience values about technologies or services may or may not give you information about what makes technologies or services accessible to your audience. The phrase "technology adoption" + adults aged 65 and older may yield different results than "technology values" + adults aged 65 and older. These questions can also be refined to explicitly include services or be more focused on a particular product category.

As a technical writer, you may use a range of methods to conduct background research, depending on your research question and the facilities you have available to you. For example, your organization may have a technical library or you may have access to a relevant archive or database. You will be also able to access a range of scholarly articles, trade publications, and databases through your local or university library. Consider also using available market research and government databases to help you begin to develop an understanding of the users and user groups.

Developing key search terms can be challenging. Start with the user group itself as well as other names for the user group. For adults aged 65 and over, you may want to use search terms connected to the generations that age group represents, e.g., Baby Boomer Generation or Silent Generation). As you are reading, be mindful of search terms that are emerging. Norman's article uses the terms "elderly users" and "seniors," which may also be helpful in developing additional search terms.

Technical writers also need to be mindful of evolving language practices. It is important to get feedback from audiences about how they want to be identified. Style guides including *Chicago Manual of Style, Associated Press Stylebook,* and *Conscious Style Guide* offer specific guidance for talking about groups and using sensitive language for describing the identity for groups. Organizations like the National Association of Black Journalists and National Association of Hispanic Journalists also offer guidance when it is not possible to talk to a group directly. Technical writers should be cognizant of respectful language and possible conflicting opinions about identity from multiple communities, even within the user group.

Conducting background research will help you, as a technical writer, focus your search terms and questions. If your topic is about a specific technology, like virtual assistant technologies, you should also look for studies that have tested these technologies with a variety of audiences and compare to what you have found that is specific to your audience. As you analyze previous studies, you can start making connections and synthesizing your findings in order to narrow your search terms, identify other key areas of background research, and proceed to the next phase.

Conducting background research is a complex process. You'll want to organize the research effectively so you can use it to help you move to the next stage. The deliverables you develop at this stage should be chosen and designed to best support the next stages of research and product development. A literature review matrix is particularly helpful in identifying gaps in the literature and studies that need to be updated. It can provide a better snapshot than an annotated bibliography and may be easier to update.

Use a spreadsheet to document the following items:

- Citation information
- Methods used in the study
- Sampling – who was included in the study, how were they recruited?
- Major findings
- Relevance to the project
- Relevance to the user group
- What can/should be updated about the study?

You can list other categories, depending on your research questions. For example, the technology or product or service studied alongside the user group might be really important and might require its own column.

Stage One is an opportunity to explain studies that would be really relevant to this group of users if only they were more recent. Use the matrix to identify what needs to be updated and why. For example, if you were conducting research about readable font choices for older adults, a 20-year-old study (e.g., Bernard et al.) that addresses both the user group and the product feature may be relevant to the project, but further research would be required to test fonts using current screen technology and mobile devices. This stage is also an opportunity to analyze recent studies to identify gaps that have not been addressed in the research thus far and thus studies that would benefit this user group. These gaps in the research—studies that need to be updated and data that still needs to be collected—should connect to the recommendations you make for the next steps for research.

When you are finished conducting background research or when you have collected enough background research to identify important gaps for additional study, you can proceed to the next phase. Keep in mind that conducting background research is an ongoing process that you will do throughout a project, but make sure you are critically analyzing your background research regularly so you can identify opportunities to move to other stages of research.

Document your work in an analytic memo. In your memo, make sure you include a summary of your findings; a description of your background research method, including keywords, research questions, and databases; and recommendations for next steps. You should identify gaps in the research, prioritize the next steps to addressing them, and explain how addressing those gaps will benefit the user group and the overall design process. At this time, you should also start documentation that will be used to define user groups—including journey maps, personas, use cases, and scenarios—as relevant. You will refine these documents throughout the process as you gather additional information.

Research Stage Two: Primary Research (Feedback, Observation, Gathering Information from Users)

In the background stage, you've identified gaps for additional research as well as research that needs to be updated. These gaps may includde missing information about the users. In Stage Two, you have a variety of primary research methods available so you can observe, evaluate, and gather feedback.

In Research Stage Two, you want to find out more about user goals and context of use. These are two areas that require primary research or gathering new data. In this stage you will want to design an approach that includes talking with users about their goals, direct analysis of the digital and physical context of use, and observations of how users approach the physical and digital spaces of use.

User goals and context of use are overlapping areas of inquiry but address them separately before comparing data. By addressing them separately, you can better understand the context of use, instead of relying on the user's perspective on the context of use.

Technical writers want to talk to users directly about their goals, usually through interviews or focus groups. To find out about user goals, you'll get better information from users if you practice effective interview techniques that draw the users to tell stories about using products or their experiences (Portigal). While users might have particular goals in mind, getting this additional detail can often help users think through their goals and provide richer data about their likes, dislikes, values, and interests. Focus groups allow users to riff off each other's answers, which may be helpful for your study.

You can also use surveys and questionnaires at this stage, which can be effectively combined with methods that help draw out user stories. For example, surveys and questionnaires can help you gather information at scale, but you'll want to gather enough data directly from users in order to write effective questions. They can also be used to gather demographic information or easily answered closed-ended questions that do not require a lot of writing.

St.Amant explains context of use as "the environments in which individuals employ an item" ("Of Scripts and Prototypes" 115). He describes context of use as including physical and digital settings and cultural values and expectations. St.Amant is writing about products created for a global market, which must consider a wide range of localization factors. Beyond localization concerns, context of use is an important consideration. Consider, for example, medical devices designed for lab environments or for assisted use that need to be used at home and sometimes without assistance ("Cultural Context of Care").

Context of use involves physical, social, and cultural environments and factors (Hackos and Redish 93-97). If you can narrow the environment of use (e.g., a particular space), you can observe users in that space. Early on, you may want to observe without any guidance to best capture how users approach tasks in that environment. More focused testing (Stage 3) should be designed from what you know about how users usually approach tasks in their environment. You also want to observe as much as you can about the context of use and try to make initial observations without disrupting regular activity or directing users in any way.

User goals do not always align with the use of a single product, service, or system. In fact, they rarely do. Further, context of use is rarely what designers have imagined. To better understand the importance of both user goals and context of use, consider the following example: One goal of college students is to graduate on time. In order for an institution to understand whether they are properly supporting this goal, it would first need to acknowledge the range of systems (e.g., course registration systems), documentation, and processes (course scheduling procedures, advising procedures, course rotations, etc.). These systems are affected by context of use. Do they require students to be on campus? What are the hours of availability? Can they be accessed effectively through mobile technology? Does documentation (e.g., instructions) assume access through a laptop instead of a mobile device? Finally, what is the starting point and how is that communicated? There are several stakeholders in the goal for students to graduate on time, and an institution will need to understand several factors related to context of use to effectively support this student goal.

Write a summary and recommendation memo or report for each study you conduct at this stage. If you conduct multiple studies, write an additional report that synthesizes the findings across the studies and makes recommendations for next steps. In your recommendation report, you'll note the findings of your research, your methods, and how the context of use and user goals could affect the overall design. Continue refining user documentation, such as personas and journey maps. You are ready to move on from Phase Two when you have research questions that you can test with audiences (usability testing).

Research Stage Three: Testing

The first two stages of UX research should take place early in the technical writing process. You are ready to start testing your product or service (e.g., content or a website) when you have gathered enough information about the user group and the context of use to develop something to test. Uusability testing is not the end of the technical writing process; rather, technical writers test their work with audiences when they have enough information to develop an effective prototype to test but are not so far along that they cannot make substantive changes based on user feedback.

Usability testing assesses how a user interacts with a particular feature of a product. For example, you might test how they interpret content, how they complete an activity, or how they navigate a website. You may go through multiple rounds of testing as well as gathering additional information from users as the prototype (e.g., a draft of content, wireframe, etc.) progresses. Testing after the design is complete makes it difficult to implement changes that can support users.

Card sorting is an early stage method you can use, though some would not classify it as a true usability test since the user is not interacting directly with a product. This activity focuses on user preferences in relation to the product but should be conducted early in the design process. The iterative nature of the three stages means that you can conduct activities in any stage as needed for the project and return to other stages. You can also combine a card sorting activity with your Stage Two interviews. The stages are flexible and your decisions should be based on using participants' time efficiently as well as having enough information about them and their context of use to make activities like card sorting effective.

Card sorting can be conducted in Stage Two or early in Stage Three, but placing it in Stage Three illustrates card sorting and think aloud protocols as complementary methods that can be done at different points in the design process. If you are unfamiliar with usability testing, these two methods are a great place to start.

Card sorting is incredibly helpful for thinking through possible approaches to information architecture, content categories, and information hierarchies from a user perspective. Card sorting involves using note cards with words and phrases (either physical or using software) and allowing users to organize them as they see fit.

A think out loud, think aloud, or talk aloud protocol involves prompting a user to share their thoughts while using a product or feature. This product or feature might be navigating a website, completing an activity, or using a set of instructions. It is a very common usability test and requires careful preparation for both the user and the moderator since prompting people while they're doing something can seem "unnatural" (Barnaum). A think aloud protocol helps you understand the extent that your design addresses audience needs.

When you've completed a round of usability testing, write up your results in a memo or short report. Include enough information about your usability test design for another person to repeat it. Be sure to describe the study's limitations and recruitment strategy, which can be useful as the project moves forward and as you pursue other tests. Continue honing your user documentation deliverables, such as personas. Make recommendations for further research and any design recommendations based on your results.

Conclusion

UX research is central to technical writing work. The three phases of user experience (UX) research—background, primary, and usability testing—presented in this chapter are intended to help you see research as both a fundamental and ongoing part of technical writing. This chapter provides a foundation for building your knowledge of research methods and when to use them as a technical writer. It touches on some methods and provides a framework for incorporating others. Methodological approaches are constantly emerging. This three stage UX research process is useful for effectively incorporating and situating traditional research methods as well as testing methods. For example, it can be used to incorporate emerging methods in community-based research, which foreground relationship building and reciprocity with users (e.g., see Shivers-McNair et al.).

This chapter also touches on important topics like accessibility, disability studies, identity, inclusive language practices, and inclusive design because they are central to all technical writing work, including UX research. Building your knowledge in these areas is never finished. A central requirement for technical writers' ability to advocate for users to constantly build and maintain this knowledge and to continue to learn about and from users as people.

Integrating UX research practices and building reciprocal relationships with communities is essential for technical writers and design teams to effectively serve users and build inclusive and effective designs. User communities are complex and finding out about them and listening to them requires time and commitment and reciprocity (Cardinal et al.; Smith et al.).

This work does not always easily align with corporate structures. Smith et al describe reciprocity as working with a community, for example helping with technology questions or providing other expert knowledge and engaging the community with your findings. Acharya's integrative literature review shows usability studies, guidelines and methods are evolving toward addressing social justice concerns and not just corporate concerns. He argues that "usability is not limited to what makes a product expedient to use, but also considers how the product can play a key role in improving peoples' lives" (Acharya 18). Technical writers will need to rethink metrics like speed and argue that the quality of a product or service is best measured in how it enhances the quality of life for users.

WORKS CITED

Acharya, Keshab Raj. "Promoting Social Justice Through Usability in Technical Communication: An Integrative Literature Review." *Technical Communication*, vol. 69, no. 1, 2022, pp. 6-26.

Barnum, Carol M. *Usability Testing Essentials: Ready, Set . . . Test!*. Morgan Kaufmann, 2020.

Broussard, Meredith. *More Than a Glitch: Confronting Race, Gender, and Ability Bias in Tech*. MIT Press, 2023.

Cardinal, Alison, Laura Gonzales, and Emma J. Rose. "Language As Participation: Multilingual User Experience Design." *Proceedings of the 38th ACM International Conference on Design of Communication*, 2020.

Dorpenyo, Isidore Kafui. "Risky Election, Vulnerable Technology: Localizing Biometric Use in Elections For The Sake Of Justice." *Technical Communication Quarterly*, vol. 28, no. 4, 2019, pp. 361-375.

Gonzales, Laura. *Designing Multilingual Experiences in Technical Communication*. Utah State UP, 2022.

Hackos, JoAnn T., and Janice Redish. *User And Task Analysis for Interface Design*. Wiley, 1998.

Hamraie, Aimi. *Building Access: Universal Design And The Politics Of Disability*. U of Minnesota Press, 2017.

Hamraie, Aimi. "Universal Design and The Problem Of "Post-Disability" Ideology." *Design and Culture*, vol. 8, no. 3, 2016, pp. 285-309.

Huntsman, Sherena. "Addressing Workplace Accessibility Practices Through Technical Communication Research Methods: One Size Does Not Fit All." *IEEE Transactions on Professional Communication*, vol. 64, no. 3, 2021, pp. 221-234.

Johnson, Robert R. "Audience Involved: Toward A Participatory Model of Writing." *Computers and Composition*, vol. 14, no. 3, 1997, pp. 361-376.

Jones, Natasha N. "The Technical Communicator as Advocate: Integrating A Social Justice Approach In Technical Communication." *Journal of Technical Writing and Communication*, vol. 46, no. 3, 2016, pp. 342-361.

Jones, Natasha N., and Miriam F. Williams. "Technologies Of Disenfranchisement: Literacy Tests and Black Voters In The US From 1890 To 1965." *Technical Communication*, vol. 65, no. 4, 2018, pp. 371-386.

Leydens, Jon A., and Juan C. Lucena. *Engineering Justice: Transforming Engineering Education and Practice.* John Wiley & Sons, 2017.

Norman, D. A. "I Wrote the Book On User-Friendly Design: What I See Today Horrifies Me." *Fast Company.* 08 May 2019. https://www.fastcompany.com/90338379/i-wrote-the-book-on-user-friendly-design-what-i-see-today-horrifies-me.

Oswal, S. K. and Palmer, Z. B. "A Critique of Disability and Accessibility Research in Technical Communication Through the Models of Emancipatory Disability Research Paradigm and Participatory Scholarship." *Assembling Critical Components: A Framework for Sustaining Technical And Professional Communication*, edited by Joanna Schreiber and Lisa Melonçon, The WAC Clearinghouse and UP of Colorado, 2022, pp. 243-268. https://doi.org/10.37514/TPC-B.2022.1381.2.09.

Rose, Emma J., and Joanna Schreiber. "User Experience and Technical Communication: Beyond Intertwining." *Journal of Technical Writing and Communication*, vol. 51, no. 4, 2021, pp. 343-349.

Portigal, Steve. *Interviewing Users: How to Uncover Compelling Insights.* Rosenfeld Media, 2013.

Sánchez, Fernando. "Racial Gerrymandering and Geographic Information Systems: Subverting The 2011 Texas District Map With Election Technologies." *Technical Communication*, vol. 65, no. 4, 2018, pp. 354-370.

Shivers-McNair, Ann, Laura Gonzales, and Tetyana Zhyvotovska. "An Intersectional Technofeminist Framework for Community-Driven Technology Innovation." *Computers and Composition*, vol. 51, 2019, pp. 43-54.

Smith, Allegra, Nupoor Ranade, and Sweta Baniya. "Local Users, Global Takeaways: Methodological Considerations for Audience Advocacy In Communication Design Research." *Proceedings of the 39th ACM International Conference on Design of Communication*, 2021.

St.Amant, Kirk. "Of Scripts and Prototypes: A Two-Part Approach to User Experience Design For International Contexts." *Technical Communication*, vol. 64, no. 2, 2017, pp. 113-125.

St.Amant, Kirk. "The Cultural Context of Care in International Communication Design: A Heuristic For Addressing Usability In International Health And Medical Communication." *Communication Design Quarterly Review*, vol. 5, no. 2, 2017, pp. 62-70.

Ward, Mark. "What do Prince Charles and Ozzy Osbourne have in common?" *BBC News.* 09 Dec. 2016, https://www.bbc.com/news/technology-37307829.

World Wide Web Consortium (WC3). "Accessibility." https://www.w3.org/standards/webdesign/accessibility. Accessed 11 May 2023.

Teacher Resources

Overview and Teaching Strategies

Audience is infused in technical writing practice and pedagogy. UX research is a formal and systemic approach to audience analysis that can be incorporated into your courses where you would normally focus on audience.

The three stages of UX research are designed to be flexible in practice but are also flexible for classroom pedagogy. For example, as a teacher you may choose to make Phase One a collaborative project where the class completes a literature review matrix together. This project can be set up with classroom activities like reading and analyzing at least one research article together and comparing audience assumptions with actual data.

UX research and the concepts from this article may be incorporated into the following topics:

- Introducing rhetorical concepts of audience and audience analysis
- Writing complex information for audiences
- Biases and assumptions about audiences
- Case studies about design failures
- Introducing concepts of usability, readability, and accessibility
- Introducing ethics and social justice approaches to technical writing
- User documentation like use cases and journey maps as project management tools

Discussion Questions

These discussions and activities can be scaled for class discussion, group activities and projects, or individual assignments.

- Choose a generation (e.g., Baby Boomers, Gen X, Millennials, or Gen Z) and write down what you know or think you know about them. Conduct background research to check your assumptions.
- Create a literature review matrix with at least 6 sources (See Stage One). Write an analytic memo or create a persona based on your findings. For information about creating personas, visit https://digital.gov/2015/04/06/using-personas-to-better-understand-customers-usa-gov-case-study/.
- Read a usability study as a class. Identify and describe the research question(s), method, and sampling approach. What remains relevant about the study and what can be updated? Consider different user groups, technologies, and contexts of use.

- Choose a campus process to evaluate. Consider processes associated with transportation, registration, financial aid, the library, or advising. Identify design decisions and how they connect to various stakeholders and audiences. What is assumed about the context of use? Design a card sorting exercise or think aloud protocol to test one of these processes.

13 "Not So Fast": Centering Your Users to Design the Right Solution

Candice Lanius and Ryan Weber

Scenario 1: You have a great idea for a new invention: an umbrella that doubles as a portable coffee maker! But before you start creating prototypes and getting startup funding from investors, it is important to think "not so fast. Does this invention solve a problem for real people?" To succeed, new products must solve people's problems and give them positive experiences.[1]

Scenario 2: Your company has an existing app that allows people to find dog sitters, and they want to add a new feature that lets people get take-out food from local restaurants delivered to their dog. Management may be tempted to assign programmers to start developing this feature right away. However, a better approach would be to say "not so fast. Do people actually want this feature? Will they actually use it?" Ideally, new product features are designed based on user research, input, and needs.

Scenario 3: Your manager strolls by your desk and says, "We need a 300-page PDF user manual that documents every feature of our website." You'll be tempted to jump in and get started. But if you can, think "not so fast. Does the audience really need a 300-page PDF? Maybe they want web help or a quick start guide instead." In some cases, your manager will have done the legwork to know that users need the long manual. But too often, technical writing projects are based on assumptions, rather than research, about what users want and need.

User-centered design (often abbreviated as UCD) is a methodology and work process where you do not make assumptions about what your audience or user needs (Lowdermilk; Norman). Instead, you "center"

---

1 This work is licensed under the Creative Commons Attribution-NonCommercial-NoDerivatives 4.0 International License (CC BY-NC-ND 4.0) and is subject to the Writing Spaces Terms of Use. To view a copy of this license, visit http://creativecommons.org/licenses/by-nc-nd/4.0/, email info@creativecommons.org, or send a letter to Creative Commons, PO Box 1866, Mountain View, CA 94042, USA. To view the Writing Spaces Terms of Use, visit http://writingspaces.org/terms-of-use.

the user at the beginning of any design, product development, or writing project to make sure it meets the needs of the person who will ultimately use it: the "end user." This chapter will introduce you to some examples of strong user-centered design and some guidance on how to center users in your design and writing processes. We clarify how user-centered design differs from usability and end with a case study that will inspire discussion and reflection on how you can use user-centered design principles in the future.

What is User-Centered Design?

People want products, services, and documents that meet their needs and that they can easily use. This sounds obvious. But designers and writers can lose sight of the people who will actually use their products or documents (people we call "users" in this chapter) when the designers/writers are in the middle of their work. There are many ways to create things that don't help users. Entrepreneurs or tech startups might create a really cool, slick, and technologically sophisticated product, only to discover that few people actually need it. Designers and companies might get attached to a product feature that users don't want and struggle to give it up because they have put so much time and energy into developing it. Designers might assume that users think like them, when instead users have totally different knowledge and expectations. Writers might create user help documents or topics that follow a company template instead of finding out how people actually use the help. None of these designers and writers are thinking about what their users actually want and need.

The focus on creating products that users actually need and can actually use is called *user-centered design*. A more formal and detailed definition of a related concept called "human-centred design" comes from the ISO (International Organization for Standardization), which sets standards for businesses around the world. According to *ISO 9241-210: Human-Centred Design for Interactive Systems*, "Human-centred design is an approach to interactive systems development that aims to make systems usable and useful by focusing on users, their needs, and requirements" (International Organization for Standardization vi). Notice that this definition emphasizes that things should be usable and useful. That is, products and documents should have good usability so users can easily and intuitively operate them (See Chapter 11 for more information on usability). But products and documents should also be useful, allowing people to complete tasks they actually need to complete. For instance, a website is usable if the menus allow users to easily find information, and it is useful if it contains information users want in the first place. User-centered design processes lead to useful products while usability testing leads to products that lack frustration during use.

User-centered design applies to the creation of products, services, and documents. In user-centered design, the needs of users take priority over the ideas and wishes of the designers, writers, and organization. People practicing user-centered design often have to set aside their own plans and egos to make sure that users come first. User-centered design starts with questions about users. Who are they? What do they want? What are their goals for using this product or document? What problems do they have that our product can solve? What experiences do they have with our product and similar products? What do they like and not like about our products or documents? Successful user-centered design requires significant knowledge about how users think, feel, and behave. It also takes a long-term perspective on users, considering their needs and behavior before, during, and after their engagement with the product document. User-centered design also considers users' long-term relationships with products and documents. When you practice user-centered design, you might have to sacrifice a pet project or a really innovative idea if it does not meet the users' needs.

User-centered design develops its perspective on users through a process that is iterative (that is, the process repeats and continues in cycles until the product truly meets user needs). This iterative process involves research and user input throughout design or writing (See Figure 13.1). The user-centered design process should begin with research and engagement with users to find out what they want and need. This research should lead to the development of product or business requirements that drive design decisions. These requirements lead designers to create product prototypes, which usually start simple and get more sophisticated as product development continues. As the business develops prototypes, it should test them with users to make sure its designs are useful and usable. If user research and testing reveal potential problems, designers should reevaluate and redesign to improve the product or document. Eventually, the business will release the product, but even then, they should provide ways to get user feedback through feedback forms or live tech support to make continuous upgrades and improvements based on user needs.

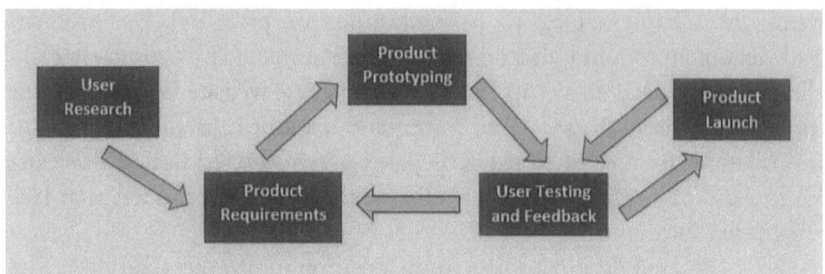

Figure 13.1: Iterative User-Centered Design Process

Often, user-centered design gets referred to by related names, like "design thinking" or "user experience" (UX). User-centered design can involve expertise from a variety of disciplines, including human-computer interaction (HCI), psychology, human factors (the study of how people perform tasks), ergonomics (the study of risk and discomfort during work), cognitive science, marketing, and technical communication. User-centered design will also involve input from many different kinds of professionals, including programmers, project managers, researchers, quality assurance professionals, graphic designers, sales people, and customer support. Because technical writers serve as user advocates who constantly remind organizations to consider and meet user needs, technical writers can be integrally involved in user-centered design.

Of course, to understand what users want, you have to engage with them in some way. A common acronym in user-centered design is GOOB, which stands for "Get Out Of The Building!" You have to leave your desk or workspace and find current or potential users so you can learn about them and let that knowledge drive your design. You want to design specifically for your users instead of for a generic user or an imagined user. Understanding users requires more than just knowing their demographics. For instance, knowing that your users tend to be 30 to 40-year-old American mothers will not give you specific enough information to design successfully for them, because you still do not know the goals, tasks, and emotions of this group. Because each product is different, you cannot always apply insights from user research about one product to another. For instance, PC and Mac users tend to be very different, even though they both use laptops. Remember too that the user for a product or document is rarely "everyone." Some products, services, and documents have a very specific, narrow set of users with specialized goals or technical knowledge, so you will need to really get to know them. Even extremely popular products and documents with large user bases (like Facebook or TikTok and their associated user help) usually have several subgroups of users with specific goals and purposes.

To truly design and write for your users, you need to learn at least the following things about them:

Goals: User goals matter most. What goal do users fulfill by using your product, service, or document? If your product or document does not help users fulfill their goals, they won't use it, simple as that. And users rarely see using your product as an end goal. For instance, few users have the goal to use TikTok. Instead, people use TikTok to achieve other goals, such as connecting with friends, building their social status, having fun, or selling products. TikTok provides a tool for them to achieve these goals. Similarly, few users have the goal of reading the white paper you wrote. Instead, that white paper gives them knowledge they can use to achieve their actual goal.

Context of Use: Find out where and when people are using your product. They may use it in a noisy, chaotic, or dangerous environment very different from the office or lab where it was designed. Products must fit their environment. For instance, if people will use your technical instructions outside, you will need to make them weather-proof. Also consider how often people use your product. Some products, like cars and Instagram, get used every day and become second nature to users. Some products, such as tax software or wedding registries, are used only in specific circumstances. And some products, such as fire extinguishers or inflatable life rafts, are used only in emergencies. To design a product well, you have to know where and when people will use it.

Emotions: Users come to products and documents with complex emotions, and they have complex emotions about products and documents. Different user goals will likely arouse different emotions. People may already be in a good mood when they get to use an Amazon gift card, and they may already be stressed or worried when they have to complete an incident report at work. People are usually already annoyed by the time they have to crack open a manual to solve their technical problem. A company's products and services may also elicit various emotions in users, like joy, relief, frustration, and anger. Part of designing for people means designing for their emotions.

Vocabulary: Designers and users don't always have the same vocabulary. When designing product interfaces and writing instructions for users, you need to know the words that users prefer so you can speak to them in their language. For instance, software developers may refer to errors by specific numbers or jargon terms that will not make sense to users. A technical writer would need to find a term that will work for the audience of a document.

Challenges: Users often face unnecessary challenges and frustrations when using products. Many of us have had the experience of a website that requires too many clicks to complete a simple task or an app that does not let you do a task you wish you could do. User-centered design professionals often call these frustrations "pain points," and by finding out where products cause users challenges, designers can improve products to make them less frustrating. Users may face challenges due to the lack of a product or document, and understanding these challenges can help you innovate new solutions. Users may also experience challenges from their own situations. Some users may experience challenges based on differences in ability, culture, language, or literacy. By knowing where your users struggle, you can create products that best meet their needs and cause them the least amount of frustration.

Again, these seem like obvious considerations, but it is easy to lose sight of users, especially in big, complex businesses where product development takes months or even years and involves hundreds of people. User-centered design is challenging. Researching users takes time and resources that companies might be reluctant to spend (even though it pays off to say "not so fast" in the long run). Not everyone knows how to do user-centered design well. Further, when a company creates a product for millions or billions of users, like Facebook or TikTok, the company needs to understand not just one user group but many.

Despite these challenges, user-centered design has many benefits. Designing products and documents that users actually want and like using increases their chances of success. User-centered design research methods also help companies avoid the costly and time-wasting process of designing products and features that people don't want. Applying user-centered design processes to technical documentation can help users better solve their problems with products and get more out of them, resulting in greater product satisfaction and fewer returns. It's no wonder that companies such as Microsoft, Apple, Facebook, Netflix, Google, Disney, Sony, and many more emphasize user-centered design. Beyond chances for increased profits and product success, user-centered design provides an ethical approach to product design that considers users as people, enhancing their satisfaction, safety, and well-being. User-centered design can also increase empathy for users within the organization.

To be truly ethical, however, user-centered design needs to consider the full range of product users. Often, companies treat white, male, cisgender, and/or able-bodied users as representatives for everyone, meaning their products do not actually meet the needs of all users. For instance, cameras are often calibrated to highlight Caucasian skin, resulting in an inferior product for non-white users. Technical instructions often rely on color for meaning without realizing that colorblind users struggle to distinguish between some colors. And many webpages present problems for users with disabilities, such as vision or hearing impairments (See Chapter 14 for more information on accessibility best practices). Accessible design accounts for the range of physical, mental, and emotional abilities of the full spectrum of users for a product, service, or written document (Dhoundiyal). Make sure your user-centered design process is also inclusive enough to meet the needs of BIPOC, LGBTQ+, and intercultural users, among others.

How do you do user-centered design?

Now that we have covered what user-centered design is, let's jump into how you can implement user-centered design in your own projects. There are many opinions on how to successfully do user-centered design. At the end of the day, however, the "right" way to do user-centered design depends on your organization and specific user group. We encourage you to adopt any of the following methods that work within your context. As long as you remember to "center the user" at all times and think "not so fast" when asked to design/write something, you can't go wrong. Here are some of the questions that you will need to focus on during your initial user research (see Chapter 12 for more information on user experience research).

Questions About Users

The process begins by identifying central goals of the overall project (i.e., the item you are creating). Determining what these goals are is a matter of answering certain user-centered questions in a particular order.

Goals: Who is your user?

Identify the audience for your document. Try to capture information about them that goes beyond marketing or demographic categories. What are their needs and expectations?

Make sure not to overlook the full range of users. We don't want a homogenous population.

Context of Use: Where is your user?

Explore the context where your user is going to be using the product or document that you create. Is it a noisy outdoor space with many distractions? Is it a quiet library room?

Write out the different tasks your user will need to complete.

Emotions are also an important aspect. Will the user be stressed or in crunch time while trying to use your product?

Tasks: How is your user going to use the product or document?

Make sure to GOOB (Get Out Of the Building) and shadow your users. You will learn so much more by watching people work and picking up on subtle things they do, both consciously and subconsciously. By watching people, you can learn the tasks they need to accomplish and the specific ways that they get work done.

Keep track of the language they are using in the natural environment; this will be the foundation for the vocabulary you use in your document.

Monitor the user for any "pain points" or frustrations where they are unable to do a task the way they expected. You may also see that users create workarounds and creative solutions to solve problems caused by your product or document.

Answering these questions involves using different approaches—or methods—for collecting information on your users in the context where they would use the item you will create for them.

METHODS

Initial user research is accomplished using three types of methods. The first method, **contextual inquiry**, involves visiting users where they will be using the product or service to gain first-hand experience. This method involves shadowing your users and even actively participating in their activity or "workflow" to deeply understand the process. For example, if you wanted to build an umbrella with a portable coffee maker described in scenario 1 of the introduction, you need to find your potential users and see how they handle their daily caffeine fix and weather mitigation. Only by going to the field, in this case their workplace or an outdoor park, can you directly see how they enjoy their coffee. By doing direct observation, you will probably find that the need to keep an umbrella directly overhead or held upside down by the handle when not in use makes it a poor candidate for doubling as a coffee maker.

The second method involves conducting **surveys** of your target user group. The questionnaire should have a mixture of closed ended questions (for standard comparisons across the group) and open-ended questions (for rich detail about problem areas). Surveys are helpful when you are unable to travel to see your users in person or you want to get input from a larger group of users (more than 20). Let's look at how this applies to scenario 2 where the dog sitting company wants to add a restaurant delivery service with food catered to dogs. While contextual inquiry may get you a few very passionate dog lovers, surveying your user group will reveal if this service is something that would be valuable for most dog owners of the app. Closed ended questions, like "How often have you wanted food delivery for your dog? Answer: Never, Sometimes, Often, Always," and open-ended questions, like "Do you have any concerns about your dog's current nutrition?" will show if implementing this service is a good idea.

The final user research method involves conducting **interviews** with individuals or hosting **focus groups** of users. The participants answer a series of open-ended questions designed to get their attitudes and opinions on a product or service. Interviews are useful for testing assumptions about what the product should even be or engaging users when it is not possible to visit them on site. For scenario 3, where you are asked to write a 300-page user manual, it could be valuable to bring users to your company where managers can directly observe the focus group and hear the truth: web help or a quick start guide work better for the user experience than a 300-page treatise. Interviews and focus groups can be very convincing proof that supports a wiser and more strategic approach to design and writing. One way to understand how all of these different elements come together and are used would be to examine an instance, or a case, involving such practices.

Case Study: See it! Click it! Fix it!

The following case study uses a service provider for municipal governments to illustrate the previous aspects of user-centered design principles and methods. "See It! Click It! Fix It!" or "SeeClickFix" was founded in 2009 as a non-emergency communication service between citizens and their local government (See https://seeclickfix.com). There are multiple ways for citizens to submit issues to their local government, including a webpage, smart phone application, or direct phone number. To submit their service request, local citizens chose one of these three options. Once they are connected, they verify their location, then select from a preset list of potential issues. If they are reporting using the website or application, they can also include a photo of the issue, title, and brief description. After the details are provided, the request is submitted to an internal system that routes the request to the appropriate local department (such as the Mayor's office or Sanitation). The department then "acknowledges," "opens," "comments," and "closes" the request. The multitude of ways that citizens are empowered to communicate with their local government on SeeClickFix showcases user-centered design principles. The company takes a user-centered design approach in several additional ways through localization (the customization of a product or service to a specific location).

First, SeeClickFix customizes their service for each city that uses the product. Customization includes branding, with things like the city logo, icons, and color schemes (See Figure 13.2). Branding creates a sense of identity and trust between the local citizens and their government, likely leading to more positive emotions about the product.

A second user-centered design principle exhibited in the customization of the SeeClickFix applications is the inclusion of request categories that match the location. Sioux Falls, for example, has lots of snow fall so their service requests reflect that reality with a "Snow Alert Information" category. St. Petersburg, on the other hand, experiences strong tropical storms, so their categories include "Hurricane Center" and "Evacuation Zones." The product considers the user's context of use. Having categories that match each location makes it easier for users to find their specific issue, but beyond that it makes the application useful as a communication tool. If users from Sioux Falls, SD logged onto the application and saw hurricane-related requests, they would think "this tool was not designed for me." Each town has categories that match their citizen's needs, wants, and desires.

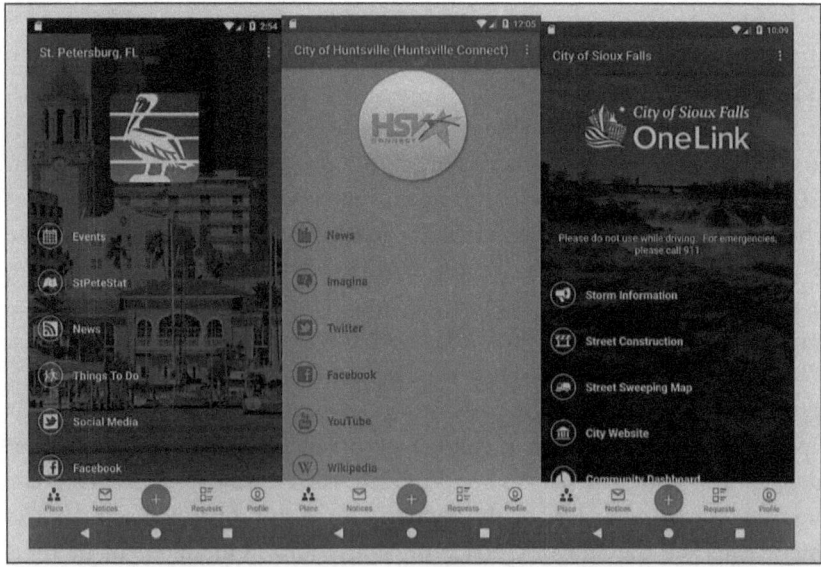

Figure 13.2: Customization for St. Petersburg, Florida, Huntsville, Alabama, and Sioux Falls, South Dakota on the SeeClickFix App. Source: Google Play. Accessed June 29, 2021.

SeeClickFix began as a product designed to meet a user need. Founder Ben Berkowitz created the product after his own frustrating experience getting graffiti removed in his neighborhood (Berkowitz & Gagnon), and the product retains that focus on helping users. Users can review the service and application directly in Google Play or the Apple Store. Based on these reviews, the SeeClickFix blog shares how they have updated their application.[2] For instance, in 2017 the product improved its color accessibility for the "7% of men and .4% of women" who "have difficulty distinguishing red from green" (Severson). SeeClickFix has also conducted user testing to "improve the functionality and SeeClickFix experience for all future iOS users" (Artsem). The web version updated its help centers in 2021 to improve usability and allow users to more quickly access information ("Changes to the CivicPlus help centers"). In this way, the product and services are focused on user needs, accessibility, and ease of use, making this an interesting case of user-centered design principles in action.

Conclusion

User-centered design can help avoid assumptions about what audiences need and help avoid creating products, such as documents, websites, infographics, etc., that individuals cannot use. By placing the user of a product at the center of the design process, user-centered design makes the processes of design, product development, or writing documents more effective and more successful. The ideas presented in this chapter introduced you to the core concepts and practices of user-centered design. By applying these ideas, you can gain a better understanding of your audience and create documents, designs, and even technologies they can more successfully use.

Works Cited

"Poor Use of SeeClickFix Input." *SeeClickFix Issues*, 14 Mar. 2018, https://seeclickfix.com/issues/4236008-poor-use-of-seeclickfix-input.

Berkowitz, Ben, and Gagnon, Jean-Paul. "SeeClickFix Empowers Citizens by Connecting Them to Their Local Governments." *Democratic Theory*, vol. 4, no. 1, Jan. 2017, pp. 121-124.

"Changes to the CivicPlus Help Centers Coming Soon." *See Click Fix Getting Started and FAQs*, 2022. https://www.seeclickfixusers.civicplus.help/hc/en-us/articles/360061732054-Changes-to-the-CivicPlus-Help-Centers-Coming-Soon-.

---

2 Since the time of writing, SeeClickFix has been updated by its parent company CivicPlus, and this blog is no longer available.

Dhoundiyal, Neha. "Inclusive Design: An Overview of Current Thinking." *UX Matters*, 26 Aug. 2019. https://www.uxmatters.com/mt/archives/2019/08/inclusive-design-an-overview-of-current-thinking.php.

International Organization for Standardization. *Ergonomics of Human-System Interaction—Part 210: Human-Centred Design for Interactive Systems*, July 2019. https://www.iso.org/committee/53372.html.

Lowdermilk, Travis. *User Centered Design*. O'Reilly Media, 2013.

Norman, Donald A. *The Design of Everyday Things: Revised and Expanded*. Basic Books, 2013.

Severson, Tucker. "Improving Color Accessibility in SeeClickFix." *Medium Product Blog*, 17 Nov. 2017. https://blog.seeclickfix.com/improving-color-accessibility-in-seeclickfix.

Figure Sources

City of Sioux Falls. https://play.google.com/store/apps/details?id=org.cityofsiouxfalls.seeclickfix. Accessed June 29, 2021.

Huntsville Connect. https://play.google.com/store/apps/details?id=com.seeclickfix.huntsvilleconnect.app. Accessed June 29, 2021.

SeeClickFix St. Pete. https://play.google.com/store/apps/details?id=com.seeclickfix.seeclickfixstpete.app . Accessed June 29, 2021.

Teacher Resources

Overview and Teaching Strategies

User-centered design helps people create better products and technical documents. Starting product designs and the writing process with a focus on users helps ensure that you will meet their needs with your finished product. Consider these ideas when implementing user-centered design:

- Prioritize the needs of users over the ideas of the designer or writer.
- Learn information about user goals, context of use, emotions, vocabulary, and challenges.
- Conduct user research using methods like contextual inquiry, interviews, focus groups, and surveys to learn about your users.
- Remember that user-centered design is an iterative process that involves users from the beginning of design and throughout the design process.

While user-centered design research can be complex and time-consuming, students can take on basic user-centered design projects and processes in the technical writing classroom. For instance, they may conduct interviews or a short survey about why users need a particular document that they are assigned to write. They might observe users complete work using a technical document (or without one, to find out how a document might help users complete their tasks). They might complete iterative document design projects that involve bringing in potential users at several stages of the writing process to help guide revision. Additionally, students might conduct user research and write a report about their findings as a written deliverable in the class.

Discussion Questions

These discussion questions are designed to accompany the SeeClickFix case study presented earlier.

1. Using the SeeClickFix website, try and find a version of the application that is close to your hometown. Based on the branding and available categories, do you think the SeeClickFix team has used user-centered design principles while localizing their service for this community?
2. Thinking about this local community, who are the target user groups of the SeeClickFix system? Make sure not to overlook anyone!
3. As a group, think of a product or service that you are all familiar with. If you were tasked with creating a list of potential user issues for this product or service, which user-centered design methods would you use?

14 Basic Approaches to Creating Accessible Documentation Projects: What Is Accessibility, and What Does It Have to do with Documentation Projects?

Cathryn Molloy

As an aspiring technical communicator, you want to learn how to take complex information and make it meaningful and useful for specific audiences and purposes.[1] Technical communicators transform knowledge about complex products so that people can use them effectively. In this way, technical communicators play a vital role in making sure that products, information, and services reach target audiences. One broad term for a range of technical communication genres is "documentation projects," or projects that provide instructions for how to use a product and/or projects that describe a technical process.

A key concern for technical communicators working on documentation projects is ensuring that the project is "accessible" or usable by everyone regardless of disability. Since technical communicators help users to comprehend and utilize specialized information and products that would otherwise be unclear, they are in the position to empower those who are too often left behind, such as those with disabilities. In this way, accessibility concerns are central concerns for the field of technical communication.

For most documentation projects you do, such as composing instructions or documenting processes, you have an "end user" (or person who'd use the product) in mind. Diligent technical communicators are very good at researching potential users in advance. Doing so allows them to make strategic decisions as they write and design a wide variety of documents. However,

---

[1] This work is licensed under the Creative Commons Attribution-NonCommercial-NoDerivatives 4.0 International License (CC BY-NC-ND 4.0) and is subject to the Writing Spaces Terms of Use. To view a copy of this license, visit http://creativecommons.org/licenses/by-nc-nd/4.0/, email info@creativecommons.org, or send a letter to Creative Commons, PO Box 1866, Mountain View, CA 94042, USA. To view the Writing Spaces Terms of Use, visit http://writingspaces.org/terms-of-use.

what can be known about potential audiences is often incomplete, making it important to consider the widest possible range of end users.

For example, let's say you're in a community-based technical communication course and you've been tasked with preparing suitable instructions for local middle school students. The assignment does not specify what process(es) or product(s) you need to document. Instead, you are asked to imagine something useful you could document for this group.

Thinking of a new skill that could be helpful for online and hybrid learning, you decide you want to assist them to learn to use the web conferencing tool Zoom. You start to imagine the average middle school student. This imagination work helps you to make good decisions as you begin to compose your project.

You reason, for instance, that most middle school students won't read above a 5th grade level, so you plan your diction (or word choice) accordingly. You imagine that some middle school students will have limited experiences using computers, so you include instructions on such basic skills as how to start up a computer and how to open a web browser. You figure that some middle school students won't know how to get their audio going, so you explain the steps involved in testing the microphone and connecting to audio on a Mac or PC.

Thinking of these preteens also gives you an insight: perhaps many would like to use headphones instead of computer audio. You go ahead and add instructions for how to do that on a Mac or PC. Another insight occurs to you: maybe they'll be on Chromebooks, tablets, or smart phones. You offer advice for these contingencies in each of your sections. You are careful and specific in your descriptions, noting that Zoom shows up as a blue icon with a camera regardless of the device used for connecting. You choose a fun, splashy color scheme for adolescents and seek out a font that appeals to kids.

Thinking about how they won't want to read anything lengthy, you decide to include lots of screenshots with dialogue bubbles to explain important parts. You're feeling pretty good about your work here. However, even with all this rich, creative thinking and planning, you've left out key parts of the middle school population—those with visual and/or hearing impairments, those with colorblindness, and those with learning disabilities. This is where accessibility in documentation projects becomes very important.

When a documentation project is **accessible**, it is usable by the widest possible group of people. In many ways, accessibility in technical communication is about basic inclusion. In other ways, thinking about accessibility in documentation projects helps you to make better written products for all users.

Let's return to the Zoom documentation project aimed at middle schoolers. When you imagine the widest possible group that could fall into your target audience, you begin to make decisions differently. Your priorities shift, and you look for ways to make sure that you aren't leaving anyone out.

Thinking of a wider range of middle school students leads you to include information on closed captioning options. You also shift the reading level down to account for those whose reading levels aren't quite to the 5th grade level yet. You start to research color decisions that take color blindness into account, and you think through how you might ensure that those with visual impairments will still be able to access the information in screengrabs.

Since technical communicators are in the business of helping users to grasp and use specialized information and products that would otherwise be too confusing or off limits, they are in the position to empower those who are too often left behind. In this way, accessibility concerns are at the heart of what is best about the field of technical communication.

What's more, far from limiting or dampening what can be done in documentation projects, accessibility considerations often lead to more vibrant and useful documents for all users.

In this chapter, you'll learn basic considerations for accessible documentation projects. The chapter will allow you to begin to approach documentation projects with fundamental accessibly in mind. In order to do so, we'll cover topics such as:

- Considering colorblindness,
- Considering dyslexia, and other conditions that alter visual abilities,
- Creating documents that are compatible with assistive devices, and
- Using alternative text for images.

Predictably, programs and tools exist to help technical communicators to create accessible documentation projects. However, it is always best to have a baseline understanding of how and why to approach accessible design in technical communication before using tools and templates.

In this chapter, therefore, you'll hear a bit about common software programs' built-in features for accessibility as well as how to use free online tools to check accessibility. Even more important than learning to use these tools, though, is learning why they are important. This chapter hopes to prove that thinking through accessibility in documentation projects enhances rather than dampens what is possible. Remember, thinking in terms of accessibility most often makes the documentation projects more usable for all users by virtue of focusing on what might be advantageous to as many users as possible. With that in mind, let's go over some basic information on disability and accessibility. After that, we'll explore some specific considerations to get you started on your journey to creating accessible documentation projects.

Disability and Accessibility Explained

Before going further with this chapter, it is necessary to take a step back and define two key concepts more carefully: **disability** and **accessibility**.

Disability is often defined in relation to the "Americans with Disabilities Act" or ADA. Signed into law in 1990 by President George H. W. Bush, the ADA made it a law that persons with disabilities be given equal opportunities to participate in mainstream American life. The ADA, in other words, made it a legal obligation to **accommodate** those with disabilities in the creation of built environments—think of elevators, wheelchair ramps, and talking crosswalk signs, and in the creation of documents and other deliverables—think of sign language interpreters that are provided during important speeches or Braille (the system of raised dots for those with visual impairments) on automatic teller machines (ATMs).

Accessibility is related to the idea of **accommodating difference**. When you make moves to alter or arrange a physical space, a written artifact, a multimodal composition, or a real time experience such that those with physical or mental disabilities can participate meaningfully, you are making moves to be accessible.

Importantly, the ADA does not give a list of conditions that are considered disabilities and, instead, defines disability broadly, as "a physical or mental impairment that substantially limits one or more major life activities, a person who has a history or record of such an impairment, or a person who is perceived by others as having such an impairment" ("What Is the Definition of Disability under the ADA? | ADA National Network"). Notice that the ADA's definition is extremely broad and open to interpretation and is designed to protect the widest group of human persons possible. It also protects a vast variety of aspects of human life, such as transportation, employment, education, health and wellness, housing, and more.

The ADA stipulates that official documents be readable via a screen reader—an assistive device that will read a document to a person with a visual impairment or other disability, such as dyslexia, that interferes with reading. Predictably, the wide and inclusive definition of disability offered in the ADA makes it difficult to offer precise statistics on how many Americans are disabled. Nonetheless, according to the National Network on Information, Guidance, and Training for the ADA, the number of disabled individuals make up between 12 and 18 percent of the population ("ADA National Network | Information, Guidance and Training on the Americans with Disabilities Act"). Undoubtedly, then, it makes not only legal sense to follow ADA requirements, but it makes good business sense to make rigorous strides to exceed legal requirements in accommodating the needs of this large group.

Considering Colorblindness

Why Consider Colorblindness?

When color is used to convey important information, people with colorblindness might miss out on that message. For example, if a background color is too close to a font color, a person with colorblindness may not even be able to tell that there is text on the page, let alone be able to read that text.

For technical communicators, the idea that text would be present, yet not readable by some audiences, is highly problematic. If there is vital information given, it's entirely lost to many audiences. Some design choices simply make it impossible to read and comprehend the content of a document. Considering colorblindness can help you to make thoughtful color and font design decisions, and such concerns are deeply related to your job as a technical communicator—to convey complicated information in a useful and user-friendly way.

Here are some things you can do to work with color in appropriate and effective ways while being inclusive of those with colorblindness and creating better documentation projects:

- Use patterns to create contrast;
- Keep color palette simple, using a maximum of three colors;
- Pay special attention to contrast; and
- Do not rely on color alone to communicate important information.

Using patterns to create contrast is a way to avoid over-relying on color differences to communicate. When color palettes have too many colors and shades, they begin to become more difficult for those with colorblindness to use. A solution is to keep color palettes simple, using no more than three colors. Equally important is making sure that the colors you are using are not so close to each other in shade that they might simply blend together for some viewers.

An easy way to make sure that your documentation project does not run the risk of having colors blend together in the audience's view is to pay special attention to contrast such that even a black and white version of what you've created will communicate the fact that two different colors are indeed different. In the same way, you want to make sure that color alone is not the means through which important information is conveyed. The following example makes these points clearer.

An Example

Let's say you've just started a new job at a brand-new coffee shop in your town called Campus Coffee. The shop has created a new menu (see Figure 14.1), and as a student in technical communication, you can't help but notice right away how inaccessible the menu is for many users.

Figure 14.1: Original Coffee Shop Menu

The background color and the font color for menu items are close in shade and tone, and nothing offsets important information other than slight color variation. In fact, as you work through your first shifts, some customers ask you to read or to clarify menu items that are difficult to read. After overhearing your new boss lamenting that she has no time to fix the issues with the menu and that she did her best to make it, you volunteer for the job. Why not, you think? It'll make a nice portfolio piece later on.

You want to use a consistent font, but you choose a simple pattern of hyphens to offset headings such that readers can easily see where cold drinks begin and hot beverages end. After doing some research, you choose a color palette that fits the overall motif of the coffee shop (warm browns and creams) yet offers you some ways to create more contrast. You go with three colors: cream, medium coffee brown, and deep teal.

You decide that it will enhance your design to use the teal for the headings, yet you rely on the line of hyphens to do the work of communicating the important information that a new section of the menu has started. Taking these small steps leads to a design you are quite proud of (see Figure 14.2).

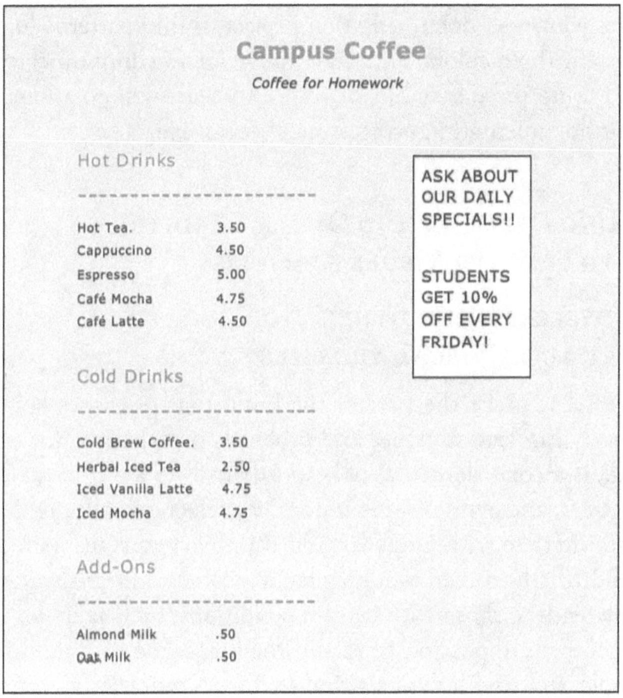

Campus Coffee
Coffee for Homework

Hot Drinks

Hot Tea.	3.50
Cappuccino	4.50
Espresso	5.00
Café Mocha	4.75
Café Latte	4.50

ASK ABOUT OUR DAILY SPECIALS!!

STUDENTS GET 10% OFF EVERY FRIDAY!

Cold Drinks

Cold Brew Coffee.	3.50
Herbal Iced Tea	2.50
Iced Vanilla Latte	4.75
Iced Mocha	4.75

Add-Ons

Almond Milk	.50
Oat Milk	.50

Figure 14.2: Revised Coffee Shop Menu

Not only are colorblind guests better served, but the menu looks better and is more usable.

How You Can Apply Considering Colorblindness to Your Schoolwork, Internships, Jobs

When you are creating documentation projects for your schoolwork, internships, and jobs, professors, internship providers, and employers will expect you to create documents that are clean and readable. In fact, today, making sure that those with colorblindness can access your documents is a baseline expectation. Like many practices related to accessibility, the advice given for how to account for colorblindness as noted—using patterns to create contrast; keeping the color palette simple; paying special attention to contrast; and avoiding using color alone to communicate important information —makes more visually pleasing and usable designs for all users. Interestingly, there are many types of colorblindness.

For an example, see the Colblindor (https://www.color-blindness.com) free color blindness simulator where you can choose from a number of options to see an image the way a person with a specific form of color blindness might. In all variations, the advice offered previously is helpful. When you approach your next documentation project, think: patterns for contrast, no more than three colors, and don't use color to communicate vital information. Taking these easy bits of advice to heart will go a long way in making your documentation projects more accessible.

Considering Dyslexia and Other Conditions that Alter Visual Abilities

Why consider dyslexia, and other conditions that alter visual abilities?

Learning disabilities that make the part of the brain that processes language work differently can lead to issues and problems deciphering documentation. Dyslexia is a condition that makes it difficult to learn to read and write words, letters, and symbols, yet it does not affect overall intelligence. Dyslexia can interfere with language and the ability to read, write, and speak. While a difficult-to-read font may require some additional concentration for many readers, those with certain conditions, such as dyslexia, may find the document impossible to read if the font is too difficult to decipher. For example, those with dyslexia find serif fonts more difficult to read—especially when the letters are close together. A study by researchers

Luz Rello and Ricardo Baeza-Yates found that *Arial italics* is especially difficult for those with dyslexia to read while fonts like **Verdana** are easier for those with dyslexia to read. For technical communicators, having words that are not readable by some audiences is a real concern, and readers with dyslexia may simply not be able to read text due to its font style. Here are some things you can do to:

- Choose fonts that work better for those with dyslexia (such as Verdana)
- Ensure that the text can easily be changed to a more readable font by the user (Zhu et al.).
- Avoid font styles that are known to be difficult for those with dyslexia to read (such as Arial Italics or other italics and serif fonts)

Choosing fonts that are easier for the user to manipulate (to change the size or look) can be a way to ensure that most users will be able to access the content of your documentation projects. While some fonts, such as Verdana, are known to be highly readable by those with dyslexia, other fonts, such as EasyReading™, have been developed specifically with dyslexic readers in mind. Even if you don't want to use a font designed for a specific disability, you can choose fonts that are easier for any user to read.

An Example

As part of a community-based technical communication course focused on the health sciences, you've been tasked with creating a brochure for a local free clinic such that a nearby high school that serves socioeconomically disadvantaged teenagers can be made aware of available services. The client has asked for paper flyers such that they can be placed on a bulletin board with other student-focused information just outside the school's cafeteria. Knowing that you will not be able to offer readers a version of the flyer where they will be able to easily manipulate the font (change it to a font that they find more readable), and thinking about users with dyslexia for whom some fonts work better than others, you create your flyer using Verdana fonts, you avoid italics, and you also include the website URL on the flyer (see Figure 14.3).

Taking these simple steps has ensured that any students with dyslexia will be able to benefit from learning more about the clinic's services. Your design is clean and easy to read; a wide variety of students at the high school will be able to get the information they need from reading it on the bulletin board.

Figure 14.3: Flyer for Health Clinic

How You Can Apply Considerations of Dyslexia to Your Schoolwork, Internships, Jobs

All of the documentation projects you work on for your schoolwork, internships, and jobs, will require you to tend to everyday communication considerations, such as audience, purpose, occasion, and exigence (or reason you are creating the document). As we learned in the section on considering colorblindness, many clients now expect you to consider disability and difference as a default when you are creating documentation projects.

Creating documents that are clean and readable by the widest group of people, including those with dyslexia for whom some fonts and styles are more difficult to read, can also be beneficial to everyone. Showing your professors, internship providers, and employers your sensitivity to difference by including accessibility considerations in your decision-making conveys to them that you are user-centered and serious about the widest group of users. Like many practices related to accessibility, the advice given for how to account for dyslexia as discussed earlier—choosing fonts that are especially readable; making digital documents revisable so that fonts can be changed or enlarged when needed; avoiding serif fonts and italics—makes more usable designs for all users.

CREATING DOCUMENTS THAT ARE COMPATIBLE WITH ASSISTIVE DEVICES

WHAT IS AN ASSISTIVE DEVICE?

Assistive device is a broad term for any device that helps a user to perform a task that would otherwise be inaccessible. Have you ever broken or sprained a leg or ankle and required crutches? If so, you've already used an assistive device. According to the World Health Organization, "assistive devices are external devices that are designed, made, or adapted to assist a person to perform a particular task" (Khasnabis et al.).

Thinking of accessible documentation projects in technical communication, assistive devices become important when considering how visually impaired or blind users will interact with your documents, yet as the previous section makes clear, if a person with dyslexia encounters a text that is in a font that is difficult for them to read, assistive devices may also be useful. Even if a person has no documented disability, cognitive styles differ, and some people simply prefer hearing to reading. There are a wide range of reasons to wish to use an assistive device, and there are also a wide range of text-to-speech assistive technologies ranging from free and simple web-based applications to sophisticated and complex cameras that will pick up and read a wide range of text.

Fortunately, if you are working on a basic documentation project in Microsoft Word, Google docs, or Adobe Pro, these programs have built-in features to help you to make sure your document will be readable via text-to-speech assistive technologies. For example, Microsoft Word and other products in the Microsoft Office suite of products have a number of built-in accessibility features. In fact, you can use preset "styles" to be reassured that the headings and subheadings you use will be readable via a screen reader. In the same way, Google Docs will automatically format documents with titles, headings, subheadings, and lists to make the work of accessible documentation projects creations easier. WIX and Canva have accessible templates for creators to use as do Adobe products. Rather than overviewing all that is out there—because there is a lot—the hope of mentioning these built-in features is that you will be inspired to seek out (and easily find) useful accessibility features in common programs.

If you are using a new program, it is worth exploring the accessibility features and templates they make available. Also noteworthy are the accessibility checkers that are now built into many programs. There is a wealth of free online tools at your disposal should you wish to check the accessibility of your documentation project. See, for example, WAVE or "web accessibly evaluation tool" at webaim.org where you can enter any URL and get a report on how accessible or inaccessible a site is.

Even without thinking about built-in accessibility features, there are some steps you can take to make sure common assistive devices can read your electronically delivered documentation:

- Use backend properties (also called "metadata" or information about a document that identify it) to tag your document with vital information for readers;
- Arrange information in a logical way;
- Make strategic arrangement decisions, using headings, lists, and meaningful hyperlinks; and
- Don't get fancy with symbols and punctuation.

One way to make your documentation project easier for a user with a visual impairment to navigate is to use backend properties or metadata—information about a document that can identify key information about that document—to deliver vital contextual information to the reader.

While some properties may be automatically generated at the backend of common programs, many also allow a user to add or edit information. For example, while Microsoft Word will automatically add the author and title, you can add a subject and keywords to enhance the usability of your document for those using screen-to-text assistive devices. Such moves give readers more context through which to make sense of your documents. As well, you can make strategic arrangement decisions, using headings, lists, and meaningful hyperlinks to improve the organization and logic of your documentation project.

Ask yourself: What are the major elements you wish to cover? Consider these worthy of their own headings. Have a big chunk of instructions to get through? Using a list will break up that information into a user-friendly flow. Want to point to a website? Describe and name the website, using a hyperlink in place of the link so that readers won't have to stumble over what will sound like gibberish to them. Such moves make your writing better organized for the reader. Finally, many text-to-speech assistive devices and programs either stumble through or simply ignore certain punctuation makes and symbols. While most will pause at periods or commas, for example many will not read a plus mark such that if you were to include something like "4 + 4 = 8", the reader would say "448." It's best to spell such things out. Taking these steps are baseline ways that you can make sure that your documentation projects are accessible to those who rely on text-to-speech assistive devices.

An Example

Imagine your internship provider, a local nonprofit focused on downtown revitalization, has asked you to take notes on how they run their annual fundraiser. Recognizing that they'll soon be hiring new staff and saying goodbye to two senior staffers, they want to document how they've done things in the past such that future leadership teams and interns will have a better idea of the steps, stages, timelines, activities, and roles that will most likely lead to a successful event.

You decide to use Microsoft Word to create a draft of this documentation. You interview the existing team, asking them to work backwards from closing out the event to all the stages and steps that went into getting them there. You go into the document properties to adjust the metadata. The screenshot in Figure 14.4 shows what you see when you go into the property summary in Word, which you can access by selecting "File," and then "Properties" from the dropdown menu.

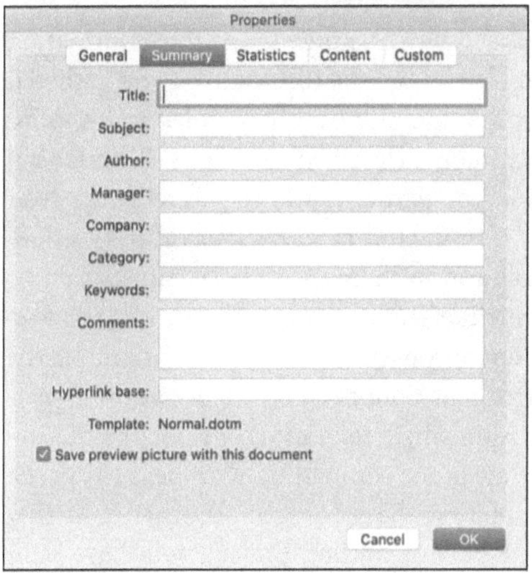

Figure 14.4: Properties Menu in MS Word

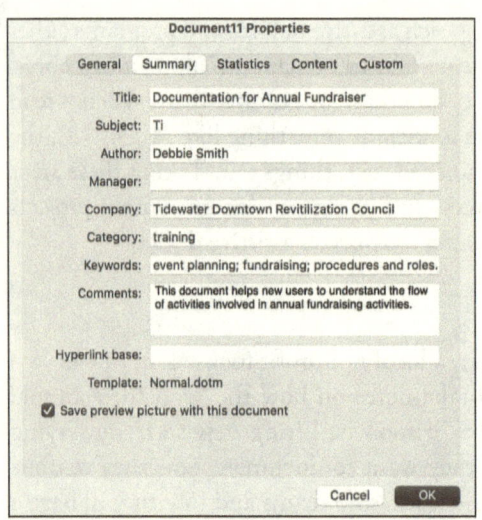

Figure 14.5: Filled In Properties Menu in MS Word

You change the author's name to your internship coordinator's name. You add a specific title, such as "Documentation for Annual Fundraiser"; keywords such as: event planning; fundraising; procedures and roles. You fill in the comments with more contextual information for users. You add "training" for the category (see Figure 14.5).

Such moves make your document easier not only for text-to-speech readers, but also for any user to get important summative information about your document.

Next, you think about a logical way to arrange all of the information you gathered. You decide that you'll arrange the information you gathered primarily via the categories: introduction and overview; roles and responsibilities; timelines; contacts; and frequently asked questions. This logical way of organizing the information you gathered allows you to create clear headings for each of these sections. In the roles and responsibilities section, you use a list to delineate each role and to share the responsibilities that go along with those roles. In the timeline section, you spell out key time-related information rather than relying on symbols or punctuation. In the contacts section, you create a "frequent donors" list where you use hyperlinks to share local business's websites where relevant. Your documentation project is now on its way to being more compatible with text-to-speech assistive technology. Again, fortunately, the software and other tools that you'll likely use for many of your documentation projects will have built in accessibility features such that doing the work of making your projects accessible will be relatively painless.

How You Can Apply This to Your Schoolwork, Internships, Jobs

When you approach documentation projects in your classes, in your internships or, eventually, in your jobs, knowing the basic logics behind how to create documentation projects that are compatible with assistive devices such as text-to-speech readers will allow you to create projects that are accessible to those with visual impairment—a population that is, according to the National Institute of Health, growing.

What's more, with the boom in popularity of audio media in things like podcasts and audio books, it's likely that even those without auditory disabilities might eventually consider using text-to-speech technologies to consume everyday documents. Either way, those tasked with creating the latest versions of popular composing tools have given a lot of time and energy to helping content creators to ensure that their documents are readable via such devices. When you approach your next documentation project, use backend properties; arrange information in a logical way; use headings, lists, and meaningful hyperlinks; and don't get fancy with symbols and punctuation. Just as in other areas of consideration for creating accessible documentation projects, taking these basic steps to ensure compatibility between your work and text-to-speech readers arguably improve the logic, look, and feel of your projects for any user.

Using Alternative Text for Images

What Is alternative text for images?

One aspect of making your documentation projects more accessible that falls under making your documents compatible with text-to-speech assistive devices and more usable for visually impaired users can be a bit trickier—using alternative text for images or "alt text." For that reason, the topic deserves its own section.

Images make powerful statements. Think about how popular memes are. They are able to communicate visual information in ways that entertain and delight users. Take, for example, the funny cat meme in Figure 14.6.

The text is an important part of the joke, yet the fact that the cat is hovering over the shoulder of a statue reading a stone book without any pages, and the fact that the cat looks as if he were reading the book, is what makes it funny. Without alt text, though, the entire message is lost to users with visual challenges.

Figures 14.6 and 14.7: Cat Memes 1 and 2

Have you ever walked into a room of friends just after the punch line to the joke has been delivered? Everyone's laughing and giving each other knowing looks. How did it make you feel? This same feeling of being left out is exactly what it feels like for those with visual impairments to find that they are missing out on vital information. Technical communicators developed the idea to add text under or attached to an image such that those with visual impairments can be "in on the joke" for things like memes. For documentation projects, the aim is not always laughter, yet the same principles apply—as a producer of content, you want to make sure that all users are in the know.

Alternative text or alt text is exactly what it sounds like it is: it's text that is used as an alternative for an image. Alt text, therefore, refers to descriptive text that replaces the image used in a project—usually a web page, and that's where the term originates—such that those with visual impairments can still benefit from images. In word processing programs such as Microsoft Word, alt text will automatically supply suggested alt text when you add an image to a document, yet these autogenerated alt text descriptions are not always precise or even accurate. Therefore, your job as a technical communicator calls on you to 1) figure out how to insert alt text in a document; and 2) compose compelling alt text such that you are satisfied that those who do not see the image will nonetheless get the message it is meant to convey.

Here are some things you can do to write vibrant alternative text for images:

- Use the "show, don't tell" principle.
- See if your alt text can pass "the yellow test."
- Gauge the purpose and aim of the image in relation to the project.

You may already be familiar with the mandate to "show, don't tell" from other genres of writing, especially creative genres. At its base, this recommendation asks you to avoid bland descriptions that lack vividness and specificity in favor of rich descriptions that allow readers to see what is in the image. It asks you to do the more difficult work of providing a snapshot of a concept or larger significance for readers. The same rule of thumb can be helpful when you are attempting to write great alt text. Show readers what's in that image by describing its details rather than just telling them in a basic way what the picture contains.

Thinking of our cat meme, for example, when you insert that image in a program like Microsoft Word, you'll get an alt text suggestion that says: "A cat sitting on top of a grass covered field." As you know from seeing the previous image, that is not the information someone would need to understand the intent of the image. That alt text tells the reader what is there, yet it doesn't use description to show the reader why that matters. I have, therefore, right clicked to add: "A cat looking over the shoulder of a stone statue of a girl with a book with the message: 'For the love of God, turn the page. You are like the slowest reader ever.'"

Another way to punch up your alt text descriptions is to see if your alt text could pass what Lee Gutkind calls "the yellow test." *Broad Street Magazine* explains that the yellow test involves using a yellow highlighter to indicate parts of a narrative where something is actually happening. The idea is that by making writing dynamic and action-oriented, it is more vivid, and readers are better engaged. Take, for example, the meme of a cat on a bed. If I insert this image, I get the suggested alt text: "A cat lying on a bed." I've used the yellow test to compose this replacement: "A kitten wedges its back against a human, looks into the camera, and raises its paws as if it were about to pounce. Text reads: 'I big scary monster raaar'" (see Figure 14.7).

You can see how trying to make the description action-oriented also does the work of making it more accurate.

Finally, a good rule of thumb when writing alt text is to consider the purpose of the image in relation to the project. Decisions on tone and word choice can often be aligned with your purpose. If your topic is serious, a formal tone is warranted. If your tone is playful or lighthearted, a lighter, fun tone is needed.

An Example

Imagine a local bridal gown seamstress has asked your technical communication class to partner with her this semester. She wants to sell patterns she's created of custom gowns she's designed from clients' mood boards as well as plans for alterations she's made to heirloom dresses to make them more modern and better fitting. Her target audience is fashion students and aspiring or even amateur seamstresses who'd like to make her gowns or who'd like to do similar work.

After collecting images from your client's personal files and scanning samples of the physical patterns she's created, you work with your classmates and professor to create a website that documents the processes used to create each custom gown. You carefully design alt text to insert under each image such that visually impaired users will have a better idea of what they might expect if they were to follow the patterns for these dresses.

For one of the gowns, you've chosen this image from the client photos the seamstress is allowing your class to use (see Figure 14.8). When you insert the image on WordPress, a plugin generates the alt text: "A person posing for a picture." However, you want to add something more descriptive. Making use of the "show, don't tell" principle, you say, "a woman stands with one hand on her hip . . ." Thinking of the yellow test, you add, "she's staring straight ahead with a lifted chin, confidently." Thinking back on your purpose—to inspire seamstresses and fashion design students to make these dresses, you add "a woman in an intricate lace, sweetheart neckline, sleeveless bridal gown . . ." Notice how the new alt text in this exercise will do a far better job of conveying the visual information the original photo is meant to communicate to potential audiences.

Figure 14.8: Wedding Dress Image

How You Can Apply This to Your Schoolwork, Internships, Jobs

When a documentation project would be truly remiss to leave out images or when an image will communicate powerful information, you will want to use them. You don't want to make it impossible for those with visual impairment to know what is on the page. Moreover, the very idea of creating alternative text or alt text dovetails beautifully with the work of a technical communicator. The same interpretative and productive skillsets that technical communicators rely on to render complex or specialized information useful and usable to non-expert audiences can come in handy when preparing appropriate and specific alternative text. In both cases, the writer must attempt to do important transformational work.

Most every documentation project will include some images, whether they are photographs, screengrabs, charts, tables, or diagrams. In all cases, these images must be interpreted for those who cannot see them—even in cases where it is possible that the data in a table will be readable via an assistive device.

Alt text is not only a matter of courtesy. Big companies, such as Target, have gotten into trouble for not providing sufficient alt text for images (Baker). Using alt text is a baseline consideration for compliance with disability standards that come out of a number of significant deliberative bodies, such as the United Nations (Ma). Thus, in your internships and jobs, you will be expected to be familiar with and to use alt text—particularly in web-based compositions. Increasingly, too, professors are expecting to see alt text used effectively in projects, and alt text is expected in video documentation projects.

When you approach your next documentation project, think: what images have I used, and do I include alt text? Am I using the "show, don't tell" principle to make my writing richly descriptive? Am I using the yellow test to make my writing dynamic and lively? And, perhaps most importantly, does my writing fit my purpose and tone?

Conclusion

As I hope this chapter has demonstrated, knowing a few basic tenets of how to approach accessibility in documentation projects is important as you decide how and when to use templates, styles, and checker tools as well as in determining their strengths and weaknesses. In the end, making documentation accessible is important as it allows all users to be included and gives them access to technical knowledge and skills.

Works Cited

"ADA National Network | Information, Guidance and Training on the Americans with Disabilities Act." https://adata.org/.

Bachmann, Christina, and Lauro Mengheri. "Dyslexia and Fonts: Is a Specific Font Useful?" *Brain Sciences*, vol. 8, no. 5, May 2018. *PubMed Central*, https://doi.org/10.3390/brainsci8050089.

"Coblis Color Blindness Simulator." *Colblindor*, https://www.color-blindness.com/coblis-color-blindness-simulator/.

Collinge, Robyn. "How to Design for Color Blindness." *GetFeedback*, 17 Jan. 2017. https://www.getfeedback.com/resources/ux/how-to-design-for-color-blindness/.

Davlin, Ann. "Color Combinations from Hell – Death Sentence for Your Designs." *DesignWebKit*, 2019, https://designwebkit.com/web-and-trends/color-combinations-hell-death-sentence-designs/.

Department of Health and Human Services Office of Communications. "What are some types of assistive devices and how are they used?" *NIH*, 24 Oct. 2018. https://www.nichd.nih.gov/health/topics/rehabtech/conditioninfo/device.

Khasnabis, Chapal, et al. "Assistive Devices." Community-Based Rehabilitation: CBR Guidelines. Geneva: World Health Organization, 2010. https://www.ncbi.nlm.nih.gov/books/NBK310951/.

"Lee Gutkind's 'The Yellow Test' and Other Advice." *BroadStreetMag*, 30 Apr. 2014. https://tinyurl.com/mu5m6tpz.

Ma, Melody. "A Brief History of Alt Text." *Medium*. 17 May 2018, https://medium.com/@melodyma/a-brief-history-of-alt-text-12cff05a59ff.

Rello, Luz and Ricardo Baeza-Yates. "Good Fonts for Dyslexia." *Proceedings of the 15th International ACM SIGACCESS Conference on Computers and Accessibility*. Association for Computing Machinery, 21 Oct. 2013, pp. 1-8. https://doi/10.1145/2513383.2513447.

"Visual Impairment, Blindness Cases in U.S. Expected To Double by 2050." *National Institutes of Health (NIH)*. 19 May 2016, https://www.nih.gov/news-events/news-releases/visual-impairment-blindness-cases-us-expected-double-2050.

What Is the Definition of Disability under the ADA? | ADA National Network." https://adata.org/faq/what-definition-disability-under-ada.

Zhu, Xinru, et al. "Analysis of Typefaces Designed for Readers with Developmental Dyslexia." *Document Analysis Systems*, edited by Xiang Bai et al., Springer International Publishing, 2020, pp. 529–43. *Springer Link*, https://doi.org/10.1007/978-3-030-57058-3_37.

Teacher Resources

Overview and Teaching Strategies

Designing documentation projects with accessibility in mind requires a sensitivity to difference and the ability to imagine the widest audience possible, including considerations of visual, cognitive, and aural disabilities. As this chapter shows, there are simple, thoughtful steps you can take to ensure that documentation projects are as accessible as possible. These steps include the following.

Considering Colorblindness

- Use patterns to create contrast.
- Keep color palette simple, using a maximum of three colors.
- Pay special attention to contrast.
- Do not rely on color alone to communicate important information.

Considering Dyslexia and Other Conditions that Alter Visual Abilities

- Choose fonts that work better for those with dyslexia (such as Verdana).
- Ensure that the text can easily be changed to a more readable font by the user (Zhu et al.).
- Avoid font styles that are known to be difficult for those with dyslexia to read (such as Arial Italics or other italics and serif fonts).

Creating Documents that are Compatible with Assistive Devices

Use backend properties (also called "metadata" or information about a document) to tag your document with vital information for readers.

- Arrange information in a logical way.
- Make strategic arrangement decisions, using headings, lists, and meaningful hyperlinks.
- Don't get fancy with symbols and punctuation.

Using Alternative Text for Images

- Use the "show, don't tell" principle.
- See if your alt text can pass "the yellow test."
- Gauge the purpose and aim of the image in relation to the project.

Gaining mastery of the basics requires students to learn the fundamental idea that creating compelling documentation projects means taking steps to accommodate the widest range of end users they can imagine. Such an approach is more inclusive, but it also makes good business sense given the rather large percentage of the population that has disabilities.

The accessibility framework described here emphasizes that rather than limiting or stifling the work of a technical communicator, though, decision-making with accessibility in mind often opens up creativity and enhances the overall quality of a documentation project. What's more, focusing on issues of accessibility is important work as the projects a technical communicator creates have the capacity to empower those who are too often left behind.

Discussion Questions

To help students explore the ideas in this entry, consider having them address—as individuals, in small groups, or as an overall class—the following questions:

1. Thinking of a recent document you created for a specific audience, purpose, and occasion, list the ways you managed to consider accessibility and the ways you would revise your composing process now that you've read this chapter.
2. How might the recommendations found in this chapter change how you approach your documentation projects?
3. What do you want to know more about in terms of accessibility? Scan the Web Accesibltiy Initiative "W3C Accessibility Standards Overview" page at https://www.w3.org/WAI/standards-guidelines/. Which topics would you like to learn more about?
4. Choose any of the main sections of this chapter and find an example of a text that does not adhere to the recommendations. Redesign the document to be more accessible.
5. This chapter makes a clear claim that considering accessibility leads to more innovative and accomplished documentation projects. To what extent do you agree or disagree with this idea, and why or why not?

Exploring these topics can help students contemplate and deliberate over how to apply the ideas presented in the chapter within the context of their own documentation projects in school and beyond.

15 Designing Multimodal Technical Instructions for Cross-Cultural Resonance Using a Culturally Inclusive Approach

Audrey G. Bennett

"The Increasingly Diverse United States of America" (https://tinyurl.com/5n7rk443) maps the evolving racial and ethnic diversity of regions nationwide (Keating and Karklis).[1] It provides compelling visual evidence that the nation is diversifying rapidly and will likely continue to do so moving forward.[2] What does this mean today for designers of technical instructions? It means that your readers are more diverse in cultural backgrounds and experiences than a decade ago and will likely continue to diversify moving forward. As access to digital media and shipped products continue to increase around the world, readers of technical instructions are becoming more global as well.

Now, we are amid an age of heightened awareness and cognizance of cross-cultural presence (e.g., consider the recent emergence of the Black Lives Matter movement) and the related rising advocacy and clarion calls for diversity, equity, access, inclusion, and justice. Within the current age of decolonization (i.e., the dismantling of a culture's domination over others), readers, the people served through our technical instructions, are culturally diverse human beings who need and want information that speaks their culture-based languages or can be accessed in a tech-mediated way that translates information into their respective languages. An example of

---

[1] This work is licensed under the Creative Commons Attribution-NonCommercial-NoDerivatives 4.0 International License (CC BY-NC-ND 4.0) and is subject to the Writing Spaces Terms of Use. To view a copy of this license, visit http://creativecommons.org/licenses/by-nc-nd/4.0/, email info@creativecommons.org, or send a letter to Creative Commons, PO Box 1866, Mountain View, CA 94042, USA. To view the Writing Spaces Terms of Use, visit http://writingspaces.org/terms-of-use.

[2] This work was supported in part by a National Science Foundation STEM+C grant #DRL-1640014. Many thanks to Principal Investigator Dr. Ron Eglash who leads the Culturally-Situated Design Tools research project.

how this cultural phenomenon is manifesting can be found in the pidgin translation of BBC's news captured as a screenshot in Figure 15.1. As a result of these cross-cultural influences on the reading of information today, technical instructions must enable cross-cultural resonance, that is, an ability to communicate effectively with people of many different cultures.

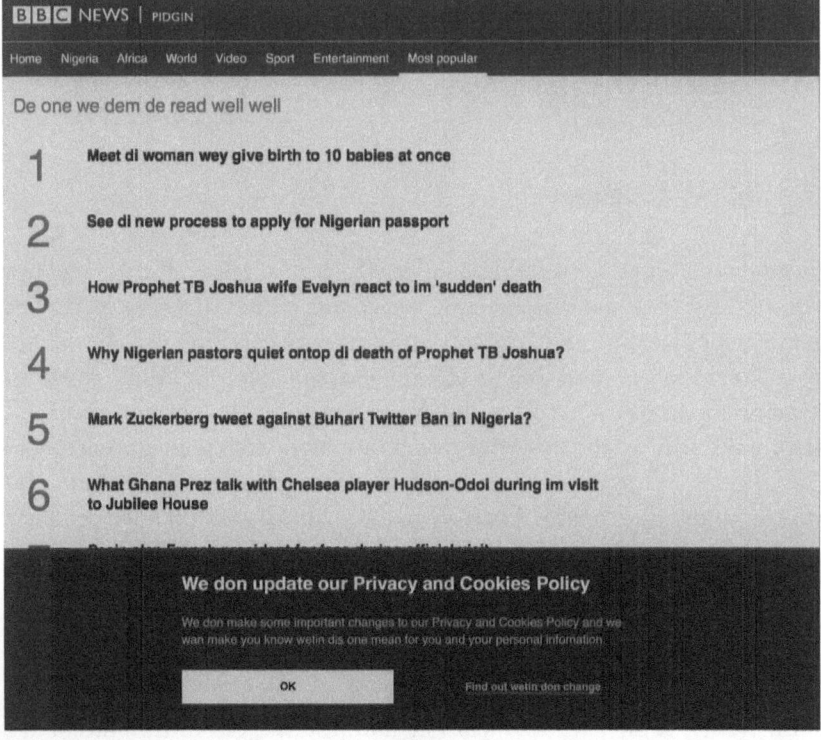

Figure 15.1 Pidgin Translation of BBC News' Most Popular Articles and its Privacy and Cookies Policy Update on Wednesday, June 9, 2021. Image Courtesy Audrey G. Bennett

When Cultural Dissonance Deters Access to Meaning

Culture, in this chapter, refers to an interaction between ethnic, racial, cognitive, gender, generational, behavioral, linguistic, political, geographic, and other identities based on heritage, choice, and environmental factors. While multimodality, that is, the integration of visual and verbal language in a sign or communicative expression (Kress) arguably has reigned as the approach to communicating to readers with multisensory and intersensory literacies or ways of knowing, it falls short in responding to cultural influences that inform a reader's sensory literacies. Clear communication occurs when there is cultural resonance between the technical information and the targeted readers. However, a variety of cultural differences among targeted readers can create dissonance in the communication of technical instructions that ultimately will deter access to or transmittance of meaning.

As readers become more global—geographically and culturally—language, aesthetic, sensory, technical, geographic, generational, socioeconomic, gender, ethnic, and other individualized social norms are some of the barriers that can emerge in the communication process and compromise the ability of the technical writer to convey information clearly. The consequences of the interference caused by cultural dissonance vary depending on the intended impact of the technical instructions. High-stakes impact may save lives; bring about life-changing cognitive, behavioral, and environmental changes; and yield equity, access, and justice. Whereas low-stakes impact may enable ease of use or functionality or understanding of how to complete a task. Whatever the intended impact, technical writers have relied heavily on multimodality to address the multisensory capabilities of readers in the communication process though with little to no consideration of cultural dissonance and the need for cultural resonance.

Designing Multimodal Technical Instructions that Resonate Across Cultures

"An Introduction to and Strategies for Multimodal Composing" (https://writingspaces.org/node/1712) explains multimodality as five ways of communicating—linguistically, visually, spatially, gesturally, and aurally—in digital and non-digital spaces that corresponds to the senses (Gagich 67). Using the term "text" to mean any communication form, from print to digital texts (e.g., movie, website, etc.), Gagich notes that these are texts that exceed the "alphabetical" and illustrates (using images) how each mode functions. For instance, one of the five images shows former First Lady of the United States Michelle Obama speaking at a secondary school in London. In the photograph, Mrs. Obama clasps her hands together, her left hand over the backside of her right hand, against her heart as she smiles at the audience. The photograph aims to clarify what the gestural mode of communication can look like. By illustrating each mode, Gagich begins to explain the rationale behind the strategic decisions that determine which form is needed when composing technical instructions. For example, Gagich specifically notes that it can aid comprehension with readers who learn differently. However, Gagich also argues more broadly that multimodality has become a standard part of daily life and, thus, must be used to meet the multiliteracy (i.e., the different ways of knowing) of diverse young readers. Gagich's perspective on multimodality focuses on addressing readers' sensory differences.

However, the contexts of communication today include more culturally diverse people, and simply using multiple modes of communicating to the different senses neither guarantees access to information nor transmittance of the intended meaning due to other barriers. One profound barrier to comprehending technical instructions is the diverse, culture-based literacies of the people to whom we are communicating. It is more likely, for all of the aforementioned reasons, that the technical instructions we design today will be read simultaneously by people from different cultures. Thus, cross-cultural resonance with a target community of diverse readers requires a different strategy when designing technical instructions. This chapter extends Gagich's modes of communicating to readers' multisensory literacies—linguistic, visual, spatial, gestural, and aural—to include their cultural literacies including language, aesthetic, sensory, technical, geographic, generational, socioeconomic, gender, racial, and ethnic perception. In the next section, I discuss how aesthetics interface with disciplines (e.g., technical communication and composition) that generate designers of technical instructions. Then, I summarize the long, cross-disciplinary, colonizing history of aesthetics and its negative impact on cultural relations in society leading to the need for a more culturally inclusive approach to the design of multimodal aesthetics that resonate cross-culturally.

Aesthetics, Technical Communication, and Composition

In "The Art of Visual Design: The Rhetoric of Aesthetics in Technical Communication," technical writer Charles Kostelnick summarizes how visual aesthetics (i.e., artistic beauty) has been a part of technical communication for centuries citing historical, fine art drawings of technologies and even technical instructions visually translated into a comic strip. Kostelnick analyzes each technical art piece's aesthetics in terms of its rhetorical impact, that is, its ability to arouse emotion and meaningful "audience engagement" (8). Kostelnick argues that visual aesthetics have the ability to engage and persuade the audience by appealing to their emotions and engendering their trust. (7) Visual aesthetics function successfully because of the "cultural knowledge [of beauty] embedded" in them that enables the "audience's" tacit recognition and understanding of the beauty reflected in their design. (6) One may incorrectly glean from this paper that there are *universal* principles for designing information. Another misleading takeaway from Kostelnick's paper may be that the sense of sight is the privileged mode in the communication process.

A broader discussion of aesthetics that shares functional value across more sensory modes can be accessed through "Reclaiming Experience: The Aesthetic and Multimodal Composition" (https://bit.ly/2RJCbl2). In this chapter, communication studies scholar Aimée Knight analyzes aesthetics as it relates to multimodal, media-based contexts of communication that facilitate meaningful 'experiences' with text beyond the sense of sight, to also include senses of sound and touch. Aesthetic sensory perception then is not limited to the visual but is embodied in a multisensory manner through the reader's lived experience in a media-infused world.

It is important to note that both Kostelnick and Knight acknowledge the importance of aesthetics theory in historical and contemporary knowledge within their related fields of technical communication and composition, respectively.

However, unlike Kostelnick, Knight, at the very least, also acknowledges that aesthetics' history and knowledge are colonized or dominated by one cultural perspective. In reference to "Western European views of aesthetics," Knight writes:

> It is important for teachers and scholars of multimodal composition to understand the story of the aesthetic . . . involves the struggle to establish the source and status of knowledge itself. I see this as a long, painful struggle filled with prized beliefs and cherished values and what gets "to count." This story has privileged certain *ways of knowing* over others, the influence of which has extended to how we teach multimodal composition today. Understanding this story helps teachers and scholars to reimagine what the aesthetic is currently in the context of multimodal composition— and what it still can be. (149)

It is important for students, too, to know the cultural history of aesthetics as students should have agency in their education and the pedagogical process. It is particularly important for designers of technical instructions who are attempting to communicate cross-culturally to understand the trauma of racial oppression that members of their targeted communities of readers who are Black, Indigenous, and People of Color (BIPOC) have endured throughout history and continue to endure in contemporary society. In the next section of this chapter, I shed light on this trauma by briefly summarizing the origin of aesthetics as beauty from the perspective of culture and the impact of this colonized definition of beauty on society leading to the systemic exclusion of BIPOC voices from cross-disciplinary knowledge systems in the West, including technical communication.

A Brief Look at the History of Aesthetics as Beauty Through a Cultural Lens

The scholarly debate on aesthetics spans many centuries and disciplines and has had a significant impact on society in ways that have informed various disciplines including design, technical communication, composition, and their evolving pedagogies, histories, and theories. Within design's canon, aesthetics has Western roots dating as far back as ancient Greek philosophy where the term aesthetics originates from the Greek word 'aesthesis,' which means sensory perception. In ancient times, aesthetics meant a kind of 'absolute' sensation. Greek philosophers Plato and Plotinus purported that aesthetics as beauty is "an ultimate value" to be pursued for its own sake; it converges with truth and goodness (Scruton 1). For instance, in The Six Enneads, Plotinus notes that "truth, beauty, and goodness are attributes of the deity" (2). In 1735 German philosopher Alexander G. Baumgarten defined aesthetics as taste or judgment of beauty in relation specifically to determining what constitutes art and the aesthetic experience of pleasure that it provides. He affirmed the assertions of the ancient Greeks, saying:

> "Beauty is the perfect perceived by the senses. Truth is the perfect perceived by reason. The good is the perfect attained by the moral will." (Baumgarten as cited in Maude 143)

Subsequently, in 1790 German philosopher Kant, at first, disagreeing with Baumgarten's trinity, later conformed to it and introduced "judgment of taste" as an experience of beauty that is based on universal subjective feelings of pleasure or displeasure (Kant).

Unfortunately, in the hands of racist scientists, Baumgarten's trinity and Kant's universal judgment of taste became a horrifying framework for uniting white supremacist aesthetics with white power politics of slavery in the 19th and 20th centuries. For instance, around the early 19th-century, a profoundly disturbing conversation among thought leaders began to coalesce around racial ranking and the inferiority of "darker races." Within this cross-disciplinary conversation, aesthetics was instrumental in laying the foundation for the rampant oppression of Black people in the 20th-century society and subsequently in the discipline of design. Paleontologist Stephen J. Gould in his book "The Mismeasure of Man" provides compelling evidence of aesthetic racism grounded in an absurd argument based on a monogenetic theory of degeneration that credited climate differences for racial differences among the descendants of Adam and Eve:

> "The most temperate climate lies between the 40th and 50th degree of latitude, and it produces the most handsome and beautiful men. It is from this climate that the ideas of the genuine color of mankind, and of the various degrees of beauty ought to be derived." The idea was that inferior non-white races could be improved in appropriate environments. Some thought leaders of the 19th-century were of a different opinion, however; they believed that racial differences were the result of the existence of separate biological species—descendants of different Adams and Eves but still with white people being the superior race." (Gould 73-74)

Gould goes on to further confront another racist argument from *"Account of the Regular Gradation in Man"* where surgeon Charles White provides the aesthetic criteria for the rank of the Caucasian (white) race as superior:

> "Where else but among Caucasians, [White] argued, can we find ... that nobly arched head, containing such a quantity of brain .... Where that variety of features, and fullness of expression; those long, flowing, graceful ringlets; that majestic beard, those rosy cheeks and coral lips? Where that ... noble gait? In what other quarter of the globe shall we find the blush that overspreads the soft features of the beautiful women of Europe, that emblem of modesty, of delicate feelings... where, except on the bosom of the European woman, two such plump and snowy white hemispheres, tipt with vermilion." (Stanton 17; as cited in Gould 73)

These prejudiced associations of aesthetics as divine beauty defined by and manifested by white Europeans set the stage for the unconscionable proliferation of aesthetic oppression of Black, Indigenous, and People of Color (BIPOC) in North America. For instance, Jim Crow-era aesthetic oppression included commercialized offensive stereotypical and exaggerated caricatures of black features that populated various forms of print media including advertisements designed to sell beauty through personal care products (e.g., soap and skin-lightening creams).

Today, the problem of aesthetic racial oppression continues systemically, for instance, in the design of urban infrastructure that segregates BIPOC communities that are economically challenged. Consequently, in "Race After Technology," sociologist Ruha Benjamin delivers an appropriately blistering and pointed critique of design as a player in contemporary aesthetic racial oppression. Specifically, she notes that "if design as a branded methodology is elevated, then other forms of generic human activity are diminished" (Benjamin 179). But is it the design method that oppresses or rather the internal or internalized racism of the designer using the approach that facilitates the creation or replication of standards of beauty grounded in centuries-old racism? Is it the mind of the designer that needs to be decolonized or freed from domination by another? Benjamin goes further to ask: ". . . would Design Thinking have helped Rosa Parks 'design' the Montgomery Bus Boycott" (176)? Building on Constanza-Schock's claim of the "universality of design as a human activity" (Costanza-Schock; cited in Benjamin 178) instead of a specialized one requiring formal training, one could argue that design thinking *did* help Rosa Parks to design the Montgomery Bus Boycott. However, as Benjamin notes further, "oppressed people and places are rarely cited for their many inventions" (Benjamin 178).

What does this mean for designers of technical instructions? When the goal is to achieve cultural resonance in the communication of information like technical instructions, the process of designing that information must change to be inclusive of reader input. In the next section, I posit that writers of technical instructions should take a culturally inclusive approach to yield cross-cultural resonance with diverse readers. I then introduce a culturally inclusive approach to design multimodal technical instructions that resonate cross-culturally by 1) representing culture appropriately from knowledge gleaned conducting primary *and* secondary research, and 2) enabling cross-cultural interpretation through interactive aesthetics.

The Culturally Inclusive Design of Multimodal Technical Instructions for Cross-Cultural Resonance

What Benjamin's critique of design calls for is a radical pivot from pre-conceived methods and principles for designing *universal* aesthetics that are deficit-based, putting the targeted reader in the position of "audience" or spectator and the designer in the position of expert or person with all of the knowledge. Primary research that is more asset-based entails designing collaboratively virtually or, preferably, in person with a diverse sampling of the targeted reader community so that the findings can be more generalizable to a larger group of readers. Virtual interaction with readers can be used in place of face-to-face engagement when there is a lack of resources or there is a real or measurable threat to one's life, like that posed during the 2020 COVID pandemic.

Engagement with the target community of diverse readers can range from conducting a focus group with a diverse sampling of readers to a participatory writing session. A focus group entails facilitating a diverse sampling of readers (typically around 25 total, with 5-8 simultaneously) giving you feedback on an iteration of the instructional text that the author has prepared in advance. Whereas a participatory writing session might entail engaging a diverse sampling of at least 35 readers in co-writing the instructional content from the beginning to end. Other approaches of primary research might include ethnographic observations of people using the technology within their cultural contexts. Ethnography is simply when we go out and observe people in their personal or work environments.

Depending on the nature or complexity of the technology, primary research efforts may also need to integrate a range of cross-disciplinary and professional and lay expertise in addition to collaboration between you and the readers over time. In inquiry-based or client-driven technical writing projects, one should solicit the input of experts from other disciplines. In inquiry-based technical writing projects, you initiate the writing project; whereas, in client-driven technical writing projects, the client initiates the technical writing project and hires you to carry it out. In inquiry-based technical writing projects, you can connect with stakeholders including experts from other disciplines through academic institutions and professional organizations. You can connect with members of the public through networking with community organizations (e.g., libraries, schools, etc.). Whereas, in client-driven technical writing projects, the client will likely assist in accessing stakeholder communities as they deem necessary.

Designing technical instructions with a culturally-inclusive approach that involves primary research is too complex to simply get it right the first time. The design process should be iterative because of the many stakeholders that may be involved. Iteration refers to something that repeats itself but improves over each cycle. Over time, the instructions will find the final form after multiple interactions between you and the different stakeholders individually and collectively throughout the writing process. Knowledge, values, and insights are exchanged and—most importantly—integrated into the evolving design. Thus, the designer and stakeholders are collaboratively controlling the decision-making that leads to the final instructional content.

Conducting secondary research of archives and open data formatively and summatively in the design process can supplement or enhance primary research efforts well though not perfectly. For instance, when used formatively, prior to primary research and to inform the interaction with culturally diverse readers, one cannot rely solely on historical, secondary data as it may be dated. Other qualitative strategies for information gathering (e.g., survey instruments) may need to be included. When used summatively to inform primary research findings on how to communicate across cultures with multimodal aesthetics, one must take precautions when secondary research findings contradict or counter primary research findings as the secondary research findings may be based on culturally-oppressive information like the aesthetics history described earlier in this chapter. To circumvent this problem, one must extend the scope of secondary research into the dark peripheries outside of design's white Western canon and access the contributions of disenfranchised BIPOC people to design, like "BIPOC Design History" (https://bipoc-designhistory.com/Index), where one will find voices left out of historical conversations on design. Within the dark peripheries, one can also find evidence of misinformation on the origins of select knowledge within the discipline's canon like that noted in "The African Origins of Swiss Design" (Bennett "The African Roots of Swiss Design") and "Follow the Golden Ratio from Africa to the Bauhaus for a Cross-Cultural Aesthetic" (Bennett "Follow the Golden Ratio from Africa to the Bauhaus for a Cross-Cultural Aesthetic for Images").

Findings from primary and secondary research inform the next phase of the culturally inclusive approach this chapter introduces. During this second phase, the designer of the technical instructions should integrate interactive aesthetics, that is tech-mediated sensory aesthetics that facilitate the reader's active and multisensory interpretation to yield cross-cultural resonance. The second phase of the culturally inclusive design process begins with the development of a plan for the aesthetic form each mode will take and how to integrate interactive aesthetics so that the instructions will communicate multimodally to the diverse readers' cultural preferences and offer opportunities for active and multisensory interpretation.

Interactive aesthetics (Bennett "Interactive Aesthetics") is a design theory that aims to bring to the forefront two things: 1) the designer's ethical responsibility in understanding the cultural impact of what they design, and 2) opportunities for remote collaboration that could facilitate stakeholder participation in the design process, including both end-users as well as other stakeholders (e.g., clients, printers, etc.). The initial idea was that with the development of more dynamic design forms (Bennett "Dynamic Interactive Aesthetics") designers could bring remotely located underserved communities into the process of designing future technologies like socially intelligent robots (Bennett "Creatively Designing Socially Intelligent Robots") to better serve the needs of underserved communities. Between the time that the interactive aesthetics (IA) paper was published and now, I've engaged in a collaborative research project centered around the collaborative and culturally inclusive design of a web-based technology called "Culturally Situated Design Tools" (see csdt.org) that teaches BIPOC youth about algorithms embedded in cultural artifacts from their cultural heritage. The next section of this chapter offers guidelines I've derived from this 20-year-plus research endeavor for translating aesthetics into interactive aesthetics starting with a plan.

Plan for Translating Aesthetics into Interactive Aesthetics Towards Cross-Cultural Resonance

An algorithm (AL-go-rith-em) is basically a set of steps for completing a task. For instance, a recipe is an algorithm. Even traveling from the airport to home has many different *algorithms* ranging from walking and taking public transportation to driving among others. Heritage algorithm refers to an algorithm that is culturally-based. (Bennett "Ethnocomputational Creativity in STEAM Education: A Cultural Framework for Generative Justice") For instance, cornrow braiding is a heritage algorithm from Africa and its diaspora.

Table 15.1 shows a plan to communicate the text 'algorithm' multimodally, culturally, and cross-culturally using interactive aesthetics. Column one shows the different modes for communicating; whereas column two shows how the text could take tangible or intangible, aesthetic form. Column three shows how the aesthetic form from column two could be extended to address any cultural dissonance that may arise in the communication process. A common feature of the culturally responsive forms listed in column two is their reliance on interactivity (i.e., active, multisensory engagement) between each mode's form and the reader.

Table 15.1 Plan for Communicating Instructions Multimodally with Interactive Aesthetics to Yield Cross-Cultural Resonance

Modes	Aesthetic form	Interactive Aesthetics
Visual	Image of code	Visual treatments and representations gleaned from primary/secondary research that reflects different cultures
		Verbal translations through roll-over images
Linguistic	Typeset text providing steps for completing a task	Language translations
Spatial	Positioning of steps in order	Positioning of steps in multiple orders of culturally-based understanding and heritage
Gestural	A computer enacts the steps to produce an action	Non-verbal communication produces the action; interaction with hyperlinked information gives way to more information (e.g., QR code, hypertext)
Aural	Sound verbalizes each step	Sound verbalizes each step in a culturally appropriate manner
		Sound technology translates a text (e.g., a word in a given language is translated into the reader's language) or image (e.g., alt-text enables people who cannot see to have access to the stories that images tell.) or aesthetic treatment (e.g., the meaning of colors, icons, symbols, etc. are translated)
Cultural	Image, sound, text, non-verbal communication, position or place	Culturally-specific representations of heritage algorithms that resonate with the culture of the reader

Table 15.2 Sample Implementation of the Plan in Table 15.1 across multiple Modes for Cross-cultural Resonance.

Aesthetic form realized	V	L	S	G	A	C
[image]	X					X

Aesthetic form realized	V	L	S	G	A	C
(star and repeat/stamp/edge flip blocks)	X		X			X
(person in hallway video)	X	X	X	X	X	X
(CSDT Anishinaabe Quilts page)	X			X	X	X
(quilt pattern)	X	X		X	X	X
(CSDT Setting up Your Script)	X			X	X	

Notes: (V=Visual, L=Linguistic, S=Spatial, G=Gestural, A=Aural, C=Cultural). Please access the alt text for each image to view a description of what is being communicated.

Implementation of the plan is where the designer applies their creative ideas. For instance, Table 15.2 shows the ways that my research team decided to implement the plan to teach BIPOC students about algorithms.

Conclusion

Today, indeed, a cross-cultural challenge presents itself when we sit down to write technical instructions. Even in the 'melting pot' state of diversity in the current United States of America and the 'flat world' afforded by evolving global access to technological innovation, we cannot assume that we possess all of the cultural knowledge necessary to communicate instructions effectively to diverse readers. Thus, this chapter proposed a culturally inclusive approach to designing multimodal technical instructions that can better resonate across cultures through the aesthetic treatment of the modes.

Works Cited

Benjamin, Ruha. *Race After Technology: Abolitionist Tools for the New Jim Code.* Polity Press, 2020.

Bennett, Audrey. "Follow the Golden Ratio from Africa to the Bauhaus for a Cross-Cultural Aesthetic for Images." *Critical Interventions*, vol. 6, no. 1, 2012, pp. 11-23.

—. "The African Roots of Swiss Design." *The Conversation*, 22 Mar. 2021, https://theconversation.com/the-african-roots-of-swiss-design-154892.

—. "Interactive Aesthetics." *Design Issues*, vol. 18, no. 3, 2002, pp. 62–69.

—. 2002. "Dynamic Interactive Aesthetics." *Journal of Design Research*, vol. 2, no. 2, 2002, pp. 91-104.

—. "Creatively Designing Socially Intelligent Robots." *International Conference on Systems, Man and Cybernetics. e-Systems and e-Man for Cybernetics in Cyberspace: October 7-10, 2001*, IEEE, 2001.

—. "Ethnocomputational Creativity in STEAM Education: A Cultural Framework for Generative Justice." *Teknocultura, Journal of Digital Culture and Social Movements* vol. 13, no. 2, 2016, pp. 587-612.

Gagich, Melanie. "Writing Spaces." *An Introduction to and Strategies for Multimodal Composing | Writing Spaces*, 1 Apr. 2020, https://writingspaces.org/node/1712.

Gould, Stephen Jay. *The Mismeasure of Man*. Penguin, 1996.

Kant, Immanuel. *Critique of Judgement*. Oxford UP, 2007.

Keating, Dan, and Laris Karklis. "Where the Country Is Becoming More Diverse." *The Washington Post*, 26 Nov. 2016, https://wwww.washingtonpost.com/graphics/national/how-diverse-is-america/.

Knight, Aimée. "Reclaiming Experience: The Aesthetic and Multimodal Composition." *Computers and Composition*, vol. 30, no. 2, 2013, pp. 146-55.

Kostelnick, Charles. "The Art of Visual Design: The Rhetoric of Aesthetics in Technical Communication." *Technical Communication*, vol. 67, no. 4, 2020, pp. 6-27.

Kress, Gunther. *Multimodality: A Social Semiotic Approach to Contemporary Communication.* Routledge, 2010.
Maude, Aylmer. *Tolstoy on Art.* Small, Maynard & Company, 1972.
Scruton, Roger. *Beauty: A Very Short Introduction.* Oxford UP, 2011.

Teacher Resources

Overview and Teaching Strategies

> *Thus is revealed the total existence of writing: a text is made of multiple writings, drawn from many cultures and entering into mutual relations of dialogue, parody, contestation, but there is one place where this multiplicity is focused and that place is the reader, not as was hitherto said, the author. We are now beginning to let ourselves be fooled no longer by the arrogant antiphrastical recriminations of good society in favor of the very thing it sets aside, ignores, smothers, or destroys; we know that to give writing its future, it is necessary to overthrow the myth: the birth of the reader must be at the cost of the death of the author.*
>
> --Roland Barthes, The Death of the Author, p. 148

Access to technologies can extend globally. However, as this chapter noted, locally within the United States of America, ethnic and cultural diversity is increasing. Thus, technical writing students, whether addressing a global or local target community, will grapple with the challenge of communicating to diverse readers. To address this challenge, instructors may supplement students' writing knowledge with training in visual communication design to acquire applied skills, strategies, and techniques to relay information aesthetically and cross-culturally. They may even integrate graphic designers as guest speakers in the course schedule to consult on how to translate the instructions visually and aesthetically.

These traditional approaches tend to keep the design of the technical instructions focused on what the author intends contrary to Barthes' perspective that the reader should be and is the focus of the communication. Since the goal essentially is to achieve cultural resonance in the communication of technical instructions, the process of designing the instructions must change to be more focused on the reader. Instead of communicating to diverse readers and users, technical communicators should pivot towards communicating with them. Teaching the next generation of professional technical communicators entails teaching them a culturally inclusive approach to technical writing that yields cross-cultural resonance with diverse readers.

The culturally inclusive approach introduced in this chapter can be used to design multimodal technical instructions that resonate cross-culturally by 1) representing culture appropriately from knowledge gleaned conducting primary and secondary research, and 2) enabling cross-cultural interpretation through interactive aesthetics. The approach aligns with a curriculum that teaches technical writing as a process of inquiry in which students engage in primary and secondary research towards developing a prototype of how instructions should be multimodally-designed with interactive aesthetics to yield cross-cultural resonance. Primary research entails engaging with the readers to gather creative insight. This interaction may take the form of surveys, questionnaires, ethnography, and other qualitative and quantitative instruments. Whereas secondary research entails accessing public archives and libraries of text and images that provide credible information that relates to the subject matter or technology or getting to know the culture of the readers.

Applied Discussion

Using the following table as a guide, independently or in small groups discuss and plan how you would go about communicating instructions for an existing technology multimodally with interactive aesthetics to yield cross-cultural resonance. To inform your development of the plan:

- Conduct secondary research to learn about the target community of readers.
- Conduct primary research to learn about how the target community of readers engage with the chosen technology.

Implement the plan you created previously by developing technical instructions that use interactive aesthetics to yield cross-cultural resonance (see Table 1). Then, use Table 2 to document the aesthetic forms realized that reflect your choices in Table 1.

Table 1

Modes	Aesthetic form	Interactive Aesthetics
Visual	Images and visual treatments	Visual treatments and representations gleaned from primary/secondary research that reflects different cultures in their motifs and designs
		Verbal translations through rollover images
Linguistic	Typeset text	Language translations

Modes	Aesthetic form	Interactive Aesthetics
Spatial	Hierarchical placement of verbal and visual elements	Positioning of information according to culturally-based understanding and heritage
Gestural	A performance	Non-verbal communication produces the action; interaction with hyperlinked information gives way to more information (e.g., QR code, hypertext)
Aural	Sound snippets	Sound verbalizes each step in a culturally appropriate manner
		Sound technology translates a text (e.g., a word in a given language is translated into the reader's language) or image (e.g., alt-text enables people who cannot see to have access to the stories that images tell.) or aesthetic treatment (e.g., the meaning of colors, icons, symbols, etc. are translated)
Cultural	Image, sound, text, non-verbal communication, position or place	Culturally-specific representations of information that resonate with the culture of the reader

Table 2

Aesthetic form realized	V	L	S	G	A	C

To evaluate your technical instructions, conduct primary research to learn about how the target community of readers engage with them and use the knowledge gained to iterate the technical instructions.

16 How to Write for Global Audiences

Birgitta Meex

Are you convinced that global companies are doing fine producing content just for one language, region, and audience?[1] And do you think that if it works in English, it'll work in other languages as well? Global content starts with translation, right? And the translator will take care of the needs of local audiences in other countries and regions. This situation is not always—or often—the case. The better you understand how such processes actually work, the more effectively you will be able to work on international projects for global clients.[2]

In this chapter, you'll hear about real-world examples and stories that say otherwise. You 'll learn why you should plan for global and write content that is "global-ready", i.e., that is suitable for international markets in terms of words, images, and design. You'll also learn how to create such content. Besides facilitating translation and localization processes, optimizing and preparing content for translation and adaptation to other languages and cultures has many advantages. The objective of the chapter is to help you understand how you, as a writer, can create texts that translators can easily work with and that can be read and understood by greater global audiences.

Writing in a translation-friendly manner not only directly impacts the quality of the translation. It boosts the quality of the source text itself (even if you aren't yet translating your content) and it affects reuse and accessibility. Texts written with translation in mind have been shown to be processed and understood more easily (also by native readers). They drive

---

[1] This work is licensed under the Creative Commons Attribution-NonCommercial-NoDerivatives 4.0 International License (CC BY-NC-ND 4.0) and is subject to the Writing Spaces Terms of Use. To view a copy of this license, visit http://creativecommons.org/licenses/by-nc-nd/4.0/, email info@creativecommons.org, or send a letter to Creative Commons, PO Box 1866, Mountain View, CA 94042, USA. To view the Writing Spaces Terms of Use, visit http://writingspaces.org/terms-of-use.

[2] I would like to thank the editors, Kirk St.Amant and Pavel Zemliansky, and two anonymous reviewers for their helpful suggestions and their valuable and constructive comments on earlier drafts of the manuscript. I am also indebted to Ken De Wachter from Flynxo for providing me with real-life examples that are used throughout this chapter to illustrate the pitfalls and strategies outlined. I also thank Annie Halseth for her meticulous editing work.

user engagement, minimize mistakes and risks, save time and money and speed up time to market. This chapter reviews potential problems that may occur from inappropriately written source texts and offers you ways to identify and solve those problems (in terms of do's and don'ts) or to avoid these problems occurring in the first place. In short, it will teach you how to become a better technical writer or technical editor.

English in the Globalized World

So why would you bother about translation and localization at all? Because today's economy is global, and companies increasingly operate across the world. A 2016 survey by Wells Fargo found that a large majority (87%) of US companies seeks business opportunities in international markets (https://newsroom.wf.com/press-release/wholesale-banking/despite-weak-global-economy-us-companies-still-turning). Tapping into new markets is an excellent way for companies to grow, to diversify and to attract foreign investments and talent. If these companies want to launch their products and services globally and in language markets other than their own, these products and services must be accompanied by information for use in the respective local language (IEC/IEEE 82079-1 Edition 2.0, 2019-05, Machinery Directive).

Besides this legal obligation, users continue to prefer to access information in their own language. Companies are therefore strongly advised to make their content locally consumable by making it linguistically and culturally appropriate for the target country and language (*locale*). Even domestic markets are made up of growing numbers of non-native speakers and their respective cultures. For example, think of the large Spanish speaking community in the U.S., whose first language is not English. Or think of the different language communities in one country adhering to different cultural value systems (e.g., Belgium, Switzerland, and Canada).

The process of addressing local users' cultural and functional expectations and needs is called *localization* (or L10n[3] if you want to be cool). One way to understand localization is as follows:

> Localization is essential to effectively addressing global markets. Being able to communicate with your target market in

---

[3] L10n is the common alphanumeric acronym for localization. 10 stands for the number of missing letters between the first letter l and the last letter n in the word.

their language, with cultural context and incorporating industry-specific aspects can mean the difference between a blockbuster product and an also-ran. (Steve Davis in Lionbridge Definitive-Guide-to-Website-Translation).

So, the U.S.-English sentence "From the driver's position on the left side of the car, you can see the sign '20 miles to destination'" would need to be converted to (localized) "From the driver's position on the *right side* of the car, you can see the sigh '*32 kilometers* to destination'". The idea is to take all items that are specific to a certain culture and revise them to reflect the expectations and experiences of a different culture.

A prerequisite for smooth localization is *internationalization* (I18n[4]), which can be thought of as follows:

> Internationalization is the process of generalizing a product so that it can handle multiple languages and cultural conventions without the need for re-design. Internationalization takes place at the level of program design and document development. (LISA[5], cited in Esselink 25)

The goal of internationalization is to remove all culture-specific items from a text so the resulting document can be understood by individuals from a range of cultural backgrounds (e.g., An English-language text any English speaker around the world could understand.) For example, if not well-designed, product information such as date formats, numbers, currency, contact or product warranty information may cause localization problems.

Different regulations in different countries are another case in point. Consider for example the famous German garbage disposal system (*Mülltrennung*). German citizens use four different color-coded recycling bins to sort and separate garbage straight at their homes: green, blue, yellow, brown, and grey ones. In many other countries, recyclable items are simply put in one larger container, without sorting them, leaving the sorting to the recyclers. When writing e.g. disposal information, you have to take into account that country and culture-specific regulations cannot be adopted in every language and for every target market. Compare in this respect the following instructions from a manual:

---

4   I18n is the common alphanumeric acronym for internationalization. 18 stands for the number of missing letters between the first letter i and the last letter n in the word.

5   The "Localisation Industry Standard Association" existed from 1990 to February 2011.

If the product is no longer required, disassemble all components and dispose plastic packaging *in the yellow bin* and packaging made of paper and cardboard *in the blue bin*.

If the product is no longer required, disassemble all components and dispose of them *in accordance with the applicable waste disposal laws in your country*.

The first instruction has not been properly internationalized and may therefore cause localization problems. It is only understandable by somebody who is familiar with the German waste disposal system, even so after translation. The second instruction, by contrast, has been made culturally neutral, which means that it has been written to serve a wide variety of markets, languages, cultures, and target audiences.

Localization and internationalization are complementary concepts, both of which are important to the overall localization process. They are two of the four interdependent processes that constitute the acronym GILT: Globalization, Internationalization, Localization and Translation (Cadieux and Esselink). Successful globalization (G11n[6]) is not possible without a firm localization strategy, which in turn relies on a proper internationalization and product design. Esperança Bielsa, a professor at the Universitat Autonoma de Barcelona in Spain, has been examining the relationship between translation and globalization for almost two decades. She sees translation as "a key mediator of global communication" (Bielsa). According to her, the role of language and translation in globalization has been systematically ignored. After all, globalization is dependent on translation for information to circulate globally. Translation is essentially about striking a balance between the local (users' language and culture respecting their habits, behaviors, frustrations, needs, and wants) and the global (brand consistency) (see e.g., Bielsa 138; Singh 86; Lionbridge; Wermuth and Meex 41). In the next section, we will see how globalization and digitalization have affected translation.

The Growth of Translation

In the previous section, we positioned translation as a crucial determinant of globalization. Digitalization and globalization have caused an exponential growth of the production of (mainly technical) content, and

---

6 G11n is the common alphanumeric acronym for globalization. stands for the number of missing letters between the first letter g and the last letter n in the word.

along with it, of translation volumes, both with respect to the number of languages involved and the number of standard pages per language. A recent survey (Meex and Straub) by tekom[7] among German manufacturing, software, and service companies revealed that on average, 11.0 different information products are created, which in turn are translated into 13.0 different languages.

According to the same survey, 46 % of technical documentation is translated into more than 10 languages. On its website, Alpha Omega Translations, a US translation company based in Alexandria VA, states that "English is the most common source language for translation in the world, but not far behind are French, German, Russian, and Italian. The most common target languages are German, French, Japanese, Spanish, and English" (https://tinyurl.com/3f8weyep). Source text quality is crucial, because issues with the source text can easily snowball into translation issues, next to usability and legal issues.

Understanding Source Text

Translation refers to "the process by which the meaning of a text written in a source language is communicated through an equivalent text in the target language" (Olvera-Lobo and Castillo-Rodríguez 34). The original language in which a text is written is called the "source language," and the original document written in that language (the document you will translate) is called the "source text." The language you wish to translate the text into is referred to as the "target language," and a text written in the target language is commonly referred to as target text. Also, the information you provide is often referred to as "content," so when we use the word content in this chapter, we are referring to the information you provide for others to read and/or translate.

In the case of technical documentation, technical writers create the source content (i.e., text to be translated), which is later translated and localized into other languages. Put differently, the work of the technical translator is preceded by that of the technical writer, who creates texts that are easy to translate and localize, or the technical editor, who revises source texts written to make them easier to translate and localize. Many aspects of content creation are relevant for later translation and localization, whether

---

7   Tekom is the largest association for technical communication in Europe. Its history dates back to 1978, the year tekom Deutschland was founded. The European umbrella association, tekom Europe, was founded in 2013 (https://www.technical-communication.org).

it be a human translation or a machine translation. Given that the vast majority of translators are native speakers of the *target* language, the quality of the source text becomes even more important.

One single mistake or inconsistency in the source text is likely to multiply by a domino effect across multiple language versions in the whole life cycle of a document, resulting in a loss of quality and resources. We can illustrate this by means of two verbs for user actions in the context of software: "Remove" and "Delete." You may be familiar with Google Drive, an application which allows users to "create, organize, discover, and share content stored on Google Drive" (https://tinyurl.com/yc4j4d8t). If you want to get rid of a doc, you see the "Remove" option combined with the trash icon in the menu on the Drive User Interface (UI). This may be somewhat confusing for you as a user (or for a translator for that matter), knowing that trash can icons are commonly used for deleting. You may ask yourself whether there is a difference between "Remove" and "Delete" and if so, what the difference is. Things become even more complicated when you notice that in other Google applications (more specifically in Google Contacts) the "Delete" option is also used with a trash can. So, what is the difference, if any? Let me explain it to you.

If you *delete* something, the information will be gone forever. If you *remove* something, it still exists in the file system, but it is not visible for the user in the file manager anymore. However, it turns out that in the case of Google, both the deleted and the removed information can be restored![8] This is confusing and inconsistent and may cause localization problems along the software life cycle. Imagine that you were supposed to translate these action verbs into German, you would probably be uncertain about whether to use "entfernen" (= remove) or "löschen" (= delete) in German. By the way, in the German language version of Google Drive, "entfernen" (= remove) is used, thus preserving the same inconsistency as in the English source text.

To sum up: when writing, always make clear to your users what will happen with their items. Educate them what the difference is. And create a consistent experience for your users! As Leah Guren, an experienced technical communication trainer and consultant from Tel Aviv, puts it: "Bad original content leads to even worse localized content!" ("Where Minimalism").

To improve the source text and to facilitate translation, proofreading may be required. Proofreading is usually performed on the target text at the final

---

8   I'm indebted to Alex Zlatkus for this example. Alex is a product designer from Boston, MA. He shared his experience and story in *UX Magazine* in 2020: https://uxmag.com/articles/ui-copy-remove-vs-delete.

stage of the translation process, but sometimes also on the source text. By definition, proofreading comes after revision and deals with the surface-level of the text only. Spelling, punctuation, grammar, and terminology are checked, whereas aspects of style and correctness of content are not. In his 2020 blog article, Stany Van Gelder, an information designer from Brussels, correctly states that "the more digitization, the higher the importance of proofreading" (https://tinyurl.com/3sa2n7yr). In the case of machine translation, adjusting the text before it is translated is called pre-editing, whereas post-editing refers to "fixing machine translation output to bring it closer to a human translation standard" (https://tinyurl.com/t8bbnpd8).

Linguistic and Cultural Competence

The international horizontal standard[9] for the development of information for use IEC/IEEE 82079-1 (Edition 2.0, 2019-05: 28) postulates that "[t]he source text should be reviewed for accuracy, consistency, and usability before it is translated." To produce quality documentation for the global market that is "locally consumable," companies need to overcome language and culture barriers and cater their content addressing the cultural and functional needs of local users and customers (Wermuth and Meex 41), i.e. analyze their native languages, values, and behaviors (Singh 86) as well as rhetorical and usability expectations (St.Amant).

To ensure smooth cultural adaptation and customization (i.e. localization), writers and translators alike need to have cultural competence next to linguistic and textual competence, as explicitly stated in international standards and competence frameworks (Meex and Straub): ISO:17100 2015 (the International Quality Standard for the translation services sector), tekom Cross-Industry Competence Framework for Technical Communication (https://tinyurl.com/3ee2paej), TecCOMFrame (https://tinyurl.com/3uu85rjy).

How to Write for Translation and Localization: Tips and Tricks

In this section you can find a selection of ten top writing rules that you can easily implement in your writing. They reflect bundles of best practices from the workplace and are commonly found in guidebooks. The selection has been compiled and slightly adapted from the following sources, which I also

---

9 A horizontal standard is a broad, general standard that is applicable across industries and product sectors.

use in my technical communication class at KU Leuven: The Global English style Guide: Writing Clear, Translatable Documentation for a Global Market (Kohl), The Definitive Guide to Website Translation (Lionbridge), Rule-based writing: English for non-native writers (tekom), and two workshops that were given by Leah Guren at the Translation Technology Summer School at KU Leuven and at tcworld in Stuttgart.

Addressing translation needs involves writing in ways that help translators better understand the meaning of the words you use, the objectives of the texts you create, and the contexts in which technical documentation will be used. Doing so often involves following certain kinds of strategies that clarify meaning and avoid confusion when writing texts—factors that enhance the usability of text for both native and non-native readers of a given language as well as for translators. Creating texts that are easier to understand and to translate often involve ten core strategies:

STRATEGY 1. KEEP IT SHORT AND SIMPLE (KISS PRINCIPLE)

The core idea of the KISS principle is to use as few and simple words as possible to get your message across. Addressing this principle involves doing the following.

- **Use short words.** Avoid nominalizations and long noun strings.
- **Use short sentences and use simple sentence constructions.** Ideally, English sentences shouldn't have more than 15-20 words. So, break up sentences and use bullets and numbered lists.
- **Use a verb-centered writing style.**

Leah Guren ("Where Minimalism") advices to write according to the principles of *minimalism* and to remove all *fluff*. She defines minimalism as "the philosophy of providing just the right information as briefly and simply as possible." With fluff she means:

- Extra and unnecessary words and information
- Complicated vocabulary
- Overly formal writing
- Vague writing
- Incorrect repetition

Remember that "[a] good technical writer reduces the word count to just the right brevity without being obscure." (https://tinyurl.com/y386wfy9) Keep also in mind that English sentences tend to get longer when they're translated. See also 10. Plan for text expansion.

Example (Lionbridge 21):

- Negative: the computer monitor sun glare
- Positive: the sun glare on the computer monitor

Strategy 2. Be Direct

Directness is important to writing in general and to writing for translation in particular because it is lean and efficient. It facilitates understanding in both the source and target languages and is easier to translate. This is especially relevant for procedural information, where you want to reach out to your audience and explain to them how they can use a product or service in a safe, efficient, and effective way. To be more direct when writing, consider doing the following things:

- Directly address your audience.
- Use the imperative for instructions and read-to-do content.
- Use the active voice in instructions and warnings. However, don't demonize the passive! Passive constructions are useful to describe automatic processes, for error messages and troubleshooting information and to give general information in tutorials. In these cases, the agent of the action is unknown or not relevant.
- Write in the simple present tense.
- Avoid modality (e.g., modal verbs and modal adverbs).
- Write positively. Avoid double negations.
- Don't be polite! It's uncommon to use the word 'please' in technical instructions.

Negative example passive voice (tekom, "Rule-Based Writing" 125):

- The software is updated on a regular basis. (user instruction)
- Positive examples active and passive voice (tekom, "Rule-Based Writing" 125):
- Update the software on a regular basis. (user instruction)
- The software is updated on a regular basis. (automatic process)
- A mistake has been made. (user error message)
- Phone calls are made by dialing the telephone number. (general information)

Example passive voice with modal verb (Kohl 42):

- Negative: The following steps should be performed to modify the Initial Contact Date.
- Positive: To modify the Initial Contact Date, perform/follow these steps.

Strategy 3. Be Specific

Specificity helps avoid confusion and error in translation by spelling out relationships between items (words, clauses, and sentences). Say exactly what you mean and don't leave out connecting devices such as conjunctions, relative pronouns, and prepositions. To be more specific when writing, focus on the following factors:

- Avoid incomplete constructions.
- Avoid unclear references across sentences.
- Don't omit functional words. Use syntactic cues: determiners (definite and indefinite articles), relative pronouns, prepositions, conjunctions, and auxiliary verbs. They make sentence structure more explicit.
- Formulate complete constructions: use conditional clauses with if ... then, use sentences containing verbs, use articles, use complete words, and always start relative clauses with a relative pronoun (tekom, "Rule-Based Writing").

Negative examples:

- If the red light stops blinking, push the START button (tekom, "Rule-Based Writing").
- Pre- and post-editing (tekom, "Rule-Based Writing").
- The software he licensed expires tomorrow (Lionbridge 21).
- Ensure the power switch is turned off (Kohl 13).

Positive examples:

- If the red light stops blinking, then push the START button (tekom, "Rule-Based Writing").
- Pre-editing and post-editing (tekom, "Rule-Based Writing").
- The software that he licensed expires tomorrow (Lionbridge 21).
- Ensure that the power switch is turned off (Kohl 13).

Strategy 4. Be Consistent

Consistency is essential to helping translators who work with source texts because it facilitates their work a lot: it is less time-consuming and results in better matches in translation memory in that the source terms match the previously translated terms in the TM. For example, the translator doesn't have to spend precious time figuring out whether a different term in the source text refers to another concept or whether it is 'just' a means to vary language, that is a different wording to say the same thing. Consistency also results in higher quality of the translation (translations are less error-prone) and in better consistency in the target texts too. To be consistent in writing, try to do the following:

- Use the same wording for the same content type (e.g., in headings).
- Avoid variation of sentences, formulations, terms, and tone for identical statements.
- Use consistent terminology. Avoid synonyms: use one term for one concept.

Positive example from Apple Support (https://support.apple.com/iphone):

- Set up your new iPhone
- Manage your Apple ID
- Customize your Home Screen

Examples for synonyms (from Philips Hue):

- conventional lighting/traditional lighting/classic lighting
- Bridge/bridge/Hue bridge/Philips Hue bridge (with and without capitalization and reference to product line or brand)

STRATEGY 5. AVOID AMBIGUOUS WORDS AND CONSTRUCTIONS

Ambiguity causes translation problems because it is not always obvious from the context which concept is intended. The context may even be missing. You can do the following things to avoid these problems when drafting texts:

- Avoid homonyms. A homonym is "a word that is spelled and pronounced like another word but is different in meaning" (Merriam-Webster). Or put differently, homonyms have the same external form but refer to different concepts. They may confuse both the user and the translator.
- Use nouns as nouns and verbs as verbs.
- Avoid ing- (gerund) constructions.
- Avoid phrasal verbs.
- Avoid jargon.
- Avoid idioms

Negative examples:

- Flying planes can be dangerous (tekom "Rule-based Writing").

- → It can be dangerous to fly planes (tekom "Rule-based Writing").
- Prostitutes appeal to pope.
- British left waffles on Falkland Islands. (former news headline)

Decontextualized sentences may be ambiguous to the extent that they become humoristic!

Example for homonym: An SME (subject matter expert) at Collibra came across the following homonym in a user interface (UI) text during revision: "Update your business terms." This sentence had been translated incorrectly into other languages. The Dutch translation "Update uw zakelijke voorwaarden (English: 'terms and conditions')" was edited out and replaced by "Update uw zakelijke termen (English: 'expressions, words')."

Example for jargon: A case study on the quality of translations of Google Help content revealed that not just highly technical terms may cause problems. Also "basic" tech terms, such as browser and URL, may confuse users (Lindsey and Swisher). It is a good idea to give an example to illustrate what the term means, for example: browsers, such as Chrome and Firefox (https://tinyurl.com/2cnub5ua). Whether or not to translate English terminology such as "wearable", "selfie", "got it" is another issue and depends on the market.

Example for idiom: Imagine, for example, the U.S. English expression "The product is a homerun." It was rendered literally as "The product quickly returns us to our home environment" in some translations but was translated metaphorically as "The product is an amazing success in others. "Be on a roll" is another example. It is highly informal and its specific, metaphorical meaning of 'to be having a successful or lucky period' cannot be derived from its constituent parts. Therefore it is likely to cause comprehension problems for non-native readers.

STRATEGY 6. USE A CLEAR STRUCTURE

By clear structure, I mean that the information is arranged in a logical order and hierarchy. A template or information model may help you draft your ideas into a meaningful collection down from the macro-level of an information product to the micro-level. A well-organized and consistent structure facilitates both the writer's and the translator's tasks. With everything in place, you can fully concentrate on drafting or translating the content and don't have to think about what to put where. As Val Swisher puts it: "when all hardware installation guides follow the same structure, the translations also follow a consistent structure" (Swisher). To structure texts more clearly, you can do the following:

- Structure your text with headings. Place the relevant information at the beginning.
- Structure the parts of a sentence according to the learning and situational logic (iconic principle). Mention the goal before the instruction. For example: "To make the text bold, click the Bold button." Avoid as soon as, prior, previously, and before. And mention the location where you have to perform the action before the action/instruction. For example: "On the ribbon, click the Underline button."
- Use standard English word order: subject – verb –object modifiers (Lionbridge 21).
- Avoid complex structures. Do not use nested clauses. Do not break the flow of a sentence with a list.

Examples:

- Negative: Perform these steps to modify the Initial Contact Date (Adapted from Kohl 42).
- Positive: To modify the Initial Contact Date, perform these steps (Kohl 42).

STRATEGY 7: AVOID MANUAL FORMATTING AND TEXT IN GRAPHICS

Manual formatting means that you format your text manually by selecting formatting options for e.g., Font and Paragraph instead of using Styles. You can also use spacings (tabs, hard returns, soft returns, page breaks) to separate text manually. Such manual formatting can cause problems for translators because it creates several segments in the translation memory. Also, text in graphics is a pain for translators because it complicates easy access to the text. To avoid such problems, consider doing the following things when creating texts:

- Do not use extra tabs, extra spaces, and hard returns.
- Do not use manual page breaks.
- Do not use hard-coded variables.
- Do not use ASCII fonts instead of Unicode.

- Do not use embedded graphics and text in graphics. Use a separate layer instead.

These are bad examples of tool usage that negatively impact translation memory systems and the final formatting of your translated content.

Strategy 8. Be Neutral

By "neutral," I mean globally understandable and accessible. Being neutral makes texts easier to translate and to localize because they lack concepts that might not work or cause problems in the target language or the target culture. Cultural-specific concepts or expressions may be completely meaningless in other languages and hence difficult to localize. To create texts that are more neutral and easy to translate you can do a number of things when writing texts.

- Avoid culturally-specific wording, concepts, imagery (e.g., hand icons, metaphor icons), maps and flags, and humor. This also includes colloquial English and regional registers.
- Use inclusive and non-evaluative language. Write gender-neutral.
- Use metric and imperial values consistently (Guren "Is Your Content").
- Use a 24-hour time designation without AM or PM and add UTC for universal reference (18:00 UTC -3) (Guren "Is Your Content").
- Use modified ISO date format (08 June 2018). Never use number-only dates (08/06/18) (Guren "Is Your Content").

Some negative examples (taken from tekom, "Rule-Based Writing" 142, 144):

- The area is as big as ten football pitches.
- Please collect the organic waste in the brown bio-waste container.
- The usage of the software is so simple that even John or Jane Doe can work with it.
- The risk assessment has to be carried out in accordance with the standard specified in VDE 0050-1.
- Laymen (instead of laypersons), old versions (instead of previous versions)

The Google case study mentioned above found that users perceived US content as too friendly. They also felt that it contained too much empathetic and reassuring language. Phrases such as the following are not meaningful to local needs:

- Don't worry, your emails aren't lost.
- It's OK, you can still do this or that.
- Remember that Google will never ask for your password.

In short, respect the conventions of the locale.

STRATEGY 9: WRITE IN TOPICS

The basic information unit in technical communication is a topic. Writing in topics means you should split up large blocks of content into smaller chunks to make the information more accessible and digestible. Writing in this way helps with translation because topics are easier to manage and translate than lengthy chapters and documents. Topics are usable in diverse contexts and content and format are separated. To write in this way you can do a number of things.

- Write once, reuse, and link many times.
- Don't write prose and big blocks of text. Use structured authoring methods instead: create small chunks of content that stand on their own and from which larger pieces of content can be built.
- Standardize your topics to divide and create your content.

Example: task, description, example, entry for glossary

STRATEGY 10: PLAN FOR TEXT EXPANSION

When content is translated from one language to another, we may witness what is termed text expansion or text contraction. Text expansion is when the target text involves more characters, words, and/or space. It often occurs in translation because languages differ with respect to the space they occupy, and it can be problematic because nowadays content is increasingly delivered and consumed on smaller screens with limited screen size. Just think of text appearing on buttons, in menu items and in check boxes. You can use certain approaches to plan for and address text expansion when writing.

Plan in advance for expansion of the translated content. If texts are translated from source English text to other Western languages such as German and French, plan for text expansion by up to 35% (The Definitive Guide to Website Translation, Lionbridge). This is especially relevant for user interface texts and navigation menus.

To implement these and other rules, companies create a term base and a style guide that can greatly help the translation process. According to a Lionbridge blog post, a translation style guide is:

"[A] set of rules for how your company presents itself textually and visually. Think of it as a guidebook for your language service provider (LSP) that includes rules about voice, writing style, sentence structure, spelling, and usage for your company." (https://tinyurl.com/227zckmj)

Developing such a translation style guide is actually the first step in the translation and localization process (https://tinyurl.com/227zckmj). In your future work as a technical writer, you may be involved in creating these important tools central to creating effective translations.

Conclusion

In today's globalized world, the demand for high quality localized content is growing. Localization entails:

- Translation of source content into one or more target languages
- Adaptation of the translated content to the needs and cultural specificity of the local market (locale)

Source text quality together with internationalization is crucial for successful localization. Source text quality serves as the foundation for translated content for different target languages, markets, and locales. As the number of target languages into which an information product is localized increases, so does the impact of the source content. However, besides international audiences also native readers benefit from translation-friendly writing.

In technical communication, quality content is lean, structured, clear, neutral, retrievable, maintainable, understandable, accessible, exchangeable, reusable, and above all consistent and translatable. Then: "consistent source content produces consistent translated content" (Swisher). By deploying a number of writing strategies, you as a writer can contribute to achieve all of these! The creation of tools such as style guides, and glossaries can help you handle your content creation. And last but not least: developing your writing for translation skills can help you succeed in the globalized job market of today.

Works Cited

Bielsa, Esperança. "Globalisation and Translation: A Theoretical Approach." *Language and Intercultural Communication*, vol. 5, no. 2, 2005, pp. 131-144.
Cadieux, Marc A., and Bert Esselink. "GILT: Globalization,

Internationalization, Localization, Translation." *Globalization Insider*, vol. 11, no. 1.5, 2002, pp. 1–5.

Esselink, Bert. *A Practical Guide to Localization*. John Benjamins, 2000.

Guren, Leah. Workshop."Where Minimalism Meets L10N. Leaner content = cleaner translation." *Translation Technology Summer School*, 3 Sept. 2019.

—. Workshop. "Is Your Content Ready for the World? Conducting a global audit." *tcworld conference*, 12-14 Nov. 2019.

"Homonym." *Merriam-Webster.com Dictionary*, Merriam-Webster, https://www.merriam-webster.com/dictionary/homonym.

IEC_IEEE_82079-1. *Preparation of information for use (instructions for use) of products – Part 1: Principles and general requirements*. IEC, 2019.

ISO 17100. Translation services -- Requirements for translation services. ISO, 2015.

Kohl, John R. *The Global English Style Guide: Writing Clear, Translatable Documentation for a Global Market*. SAS Institute Inc, 2008.

Lindsey, Jon A., and Val Swisher. "Optimizing Content for Improved Translation Effectiveness: Google Case Study". Webinar BrightTALK, 19 Sep. 2017.

Lionbridge Definitive-Guide-to-Website-Translation, 2020.

"How to Create a Translation Style Guide and Terminology Glossary." Lionbridge blog post, 5 Nov. 2016, https://www.lionbridge.com/blog/translation-localization/how-to-create-a-translation-style-guide-and-terminology-glossary/.

Machinery Directive. Directive 2006/42/EC of the European Parliament and of the Council of 17 May 2006 on machinery, and amending Directive 95/16/EC (recast). https://ec.europa.eu/growth/single-market/european-standards/harmonised-Standards/machinery_nl, 2006.

Meex, Birgitta, and Daniela Straub. "How to Bridge the Gap Between Translators and Technical Communicators?" *The Journal of Internationalization and Localization*, vol. 3, no. 2, 2016, pp. 133-151.

Olvera-Lobo, María Dolores, and Celia Castillo-Rodríguez. "Website Localisation in the Corporate Context: A Spanish Perspective." *Journal of Digital Information Management*, vol. 17, no. 1, 2019, pp. 34.

Singh, Nitish. *Localization Strategies for Global E-Business*. UP, 2012.

St.Amant, Kirk. "Writing in Global Contexts: Composing Usable Texts for Audiences from Different Cultures" *Writing Spaces: Readings on Writing*, vol. 3, edited by Dana Driscoll, Mary Stewart and Matthew Vetter, WritingSpaces.org, Parlor Press, and The WAC Clearinghouse, 2020, pp. 147-61. https://wac.colostate.edu/books/writingspaces/writingspaces3/.

Swisher, Val. "How Topic-based Authoring Benefits Translation." *madcap software*, 9 June 2016, https://www.madcapsoftware.com/blog/topic-based-authoring-benefits-translation.

TecCOMFrame. "A Joint European Academic Competence Framework and Curricula for the Training of Technical Communicators." 2017, https://www.teccom-frame.eu/.

tekom. "Competence Framework for Technical Communication: Competencies for Technical Communicators." 2015, http://competencies.technical-communication.org.

tekom. *Rule-Based Writing. English for Non-Native Writers.* 2017b.

Van Gelder, Stany. "Why is proofreading of translations a must?"
Van Gelder, Stany. "Why is proofreading of translations a must?", Connexion blog post, 30 Jan. 2020, https://www.connexion.eu/blog/why-proofreading-translations-must.

Connexion blog post, 30 Jan. 2020, https://www.connexion.eu/blog/why-proofreading-translations-must.

Wermuth, Cornelia and Birgitta Meex. "Attila Hildmann Goes International: An Explorative Study on the Localization of a German Chef's Website for a US Target Audience." *The Journal of Internationalization and Localization*, vol. 4, no. 1, 2017, pp. 40-70.

Zlatkus, Alex. "UI copy: Remove vs Delete. The difference between "remove" and "delete," and how to ensure your user understands." *UX Magazine*, June 4, 2020, https://uxmag.com/articles/ui-copy-remove-vs-delete.

Teacher Resources

Overview and Teacher Strategies

I approached this chapter with three goals in mind. First, this chapter wants to provide a context for understanding the importance of source text quality as a prerequisite for successful translation and localization in today's growing global environment. Second, given that students will likely write for audiences around the globe, they should understand why it is important to draft content that is global-ready and know which methods they can use to facilitate translatability and with it, consistency, comprehensibility and usability of technical content. Third, this chapter offers students a selection of writing rules, which they can easily implement in their own writing.

Faculty may use this chapter in teaching hands-on technical writing courses. In these courses, students need to learn and practice the core fundaments of good source content creation. In my technical communication classes at KU Leuven, I usually ask my students to first analyze and then rewrite an existing manual. For first year composition students, an instruction manual for household appliances, consumer electronics or daily medical devices is accessible and not too technical. Students discover the many inconsistencies that the manual contains and report their findings in an oral or written assignment.

This chapter is particularly suitable for teaching with the flipped classroom approach. In this blended learning approach, students read the chapter and review additional course material at home. They practice working in class. This approach has the advantage of leaving more room for collaboration, discussion, and direct feedback. Alternatively, faculty may use the chapter to introduce students to the subject and assign them tasks (analyzing, rewriting, and composing) that they should complete at home. Finally, instructors of writing may also participate in academic bilateral or multilateral writing/editing-translation projects with faculty from neighboring disciplines such as translation and usability studies to give their students experience in preparing texts for translation and localization and in working with translators. The partners in the well-established Trans-Atlantic & Pacific Project (TAPP) have been cooperating in bilateral writing-translation projects since 2001 (https://tinyurl.com/mr2ebxb9).

Discussion Questions and Sample Activities

Discussion questions that can help students further explore these ideas on language and translation could include the following:

1. Have you ever used a style guide? Which one(s)? For which purpose?
2. Consult English style guides and guidebooks for information on how to create effective source texts (for example, Kohl's Global English style guide or tekom's writing rules if you are a non-native writer of English) for more writing rules. Different students may use different style guides. Review the rules and make your own selection (one that fits your specific purpose and situation) and compilation of rules. Look for additional real-world examples to illustrate your compilation or use examples from your own writing.
3. Now it's time to attack and rewrite an existing text! Bring a manual from home. Analyze and evaluate the content. Look for unnecessary information, inconsistencies, and other shortcomings. Also describe what is good. Rewrite and optimize the content by implementing the writing rules in this chapter.
4. Consult the list of the top 50 websites from the latest version of the Web Globalization Report Card: e.g., https://www.bytelevel.com/reportcard2023/. Visit and compare the global website and one localized version of the website of a company of your choice. Present your findings to the class.
5. Build a tandem with a student in translation studies. Discuss the problems you are experiencing when writing technical texts. Listen

to the issues that your partner is having when he or she is translating a text. In the next stage, you can provide your partner with a source text that you have prepared for translation and localization.

17 Composing Technical Documents for Localized Usability in the International Context

Keshab Raj Acharya

With the international spread of business, you are now globally connected.[1] This global interconnectedness has increased your ability to understand and function as an informed member of and participant in diverse national and global cultures. To design and compose more effective texts (print, digital, multimedia, visual) for audiences from other cultures, you need to understand how cultural factors affect people's perception of your message in those cultures.

Culture matters in defining your behaviors, norms, values, and belief systems. These cultural factors affect how audiences in the international context perceive and react to the message you share with them. These cultural factors also mean that writing for international audiences is different from writing for your domestic audiences. This chapter uses different examples to illustrate these ideas. Specifically, the chapter uses these examples to demonstrate the need for writing **usable, meaningful,** and **credible messages** for audiences from other cultures. The chapter also introduces the process of document localization as a tool for creating such messages from the perspectives of multicultural audiences. The chapter concludes by providing guidelines and best practices for composing culturally usable, meaningful, and credible texts for diverse audiences in international contexts.

In your technical writing class, you might be given a number of assignments to complete individually or in teams. These assignments might include different technical documents, such as instruction manuals, memos,

---

1 This work is licensed under the Creative Commons Attribution-NonCommercial-NoDerivatives 4.0 International License (CC BY-NC-ND 4.0) and is subject to the Writing Spaces Terms of Use. To view a copy of this license, visit http://creativecommons.org/licenses/by-nc-nd/4.0/, email info@creativecommons.org, or send a letter to Creative Commons, PO Box 1866, Mountain View, CA 94042, USA. To view the Writing Spaces Terms of Use, visit http://writingspaces.org/terms-of-use.

proposals, reports, emails, and product descriptions. In composing these documents, your main goal is to have your work seen as credible or trustworthy by the intended audience. Composing such credible texts requires an in-depth understanding of:

- The audience you are writing for. (Who are they?)
- The purpose for which your audience is reading your work. (What do you want your readers to do after reading your work?)
- The context in which the audience will use your work. (Where are they when reading and using it?)
- The type of document you are composing. (What are you writing, and in what format?)

Understanding these factors can be challenging even in contexts familiar to you, but these dynamics become even more complex when writing for audiences across cultures.

Culture can be defined as "the sum total of the beliefs, rules, techniques, institutions, and artifacts that characterize human populations" (Ball and McCulloch 258). Culture is a "set of values and beliefs, or a cluster of learned behaviors" people share with other members in a particular society (Lebrón 126). In other words, culture encompasses the set of behaviors, norms, values, and belief systems—the cultural aspects held in common by a group of people in a society. As such, culture varies, and these variations affect how your audience understands and uses the documents you produce. In short, writing technical documents for audiences from other cultures involves understanding the cultural aspects of your target audiences (St.Amant 148–151).

Rhetorical Contexts of Technical Writing

Rhetoric is the process of creating messages that address the expectations and needs of your intended audience. The more you understand what your audience knows about a topic and how members of that audience discuss it, the more effectively you can compose messages that your audience can understand and use effectively—for example, using instructions to assemble a new cabinet. Technical documents need to be clear, concise, accurate, and professional. Clarity, accuracy, and precision are essential elements in composing usable, meaningful, and credible technical documents for readers from other cultures. Composing a usable or easy-to-use text is not enough; you must also think about how meaningful—that is, how the text resonates with different cultures—and how credible the text is for your target audience.

Consider a situation in which you want to use a printer in your university library to print out a class assignment. You go to the printing station where only one printer is available. As soon as you hit the print command button, you see an error message pop up on the screen, notifying that the paper tray is empty. You have ten minutes left before the library closes, and nobody is available to assist you. You look around and find reams of printing paper in a cabinet. Now, you take out a few sheets for printing, but you don't know how to load the paper into the machine. On the wall adjacent to the printer, you see a set of instructions. They read:

> If the printer has run out of paper, fill the tray back up and press the power button to go on with the print job.

If you are a native English speaker, you might not have any problem understanding and following the instructions. You know what the instructions mean, so you can perform the actions they describe and successfully complete the desired task.

Consider you are an international student with a limited command of English. Unlike the native English speaker, you are unfamiliar with some of the terms in the instructions. While reading the instructions, you become confused because you don't know what the phrases such as "run of," "back up," and "go on with" mean. You grapple with making sense of these phrases for a moment, but you have no idea that:

- "run out of" means "consume or deplete all of something"
- "fill [something] back up" means "replace the contents of a container"
- "go on with" means "continue or proceed."

As a result of not being able to accomplish the task, you leave the printing station in frustration and anger. Who do you blame? Yourself, or the unknown person who wrote these instructions? Why?

Creating Usable Text

One of the main goals of technical communication is to compose information in plain language that everyone—regardless of educational background or language proficiency—can read, understand, and use.

Composing documents in plain language means enabling your audience to "find what they need, understand what they find, and use what they understand appropriately" (Redish 163). Plain language is particularly important for international audiences, many of whom are not native speakers, or who might use a different dialect of a language, such as English, to communicate. In the printer example, it is best to remove phrasal verbs and use plain language, such as the following:

> The printer needs more paper. Refill the paper tray and press the power button to continue.

The lesson here is one of culture and comprehension. Unless your text meets the audience's expectations and needs, they will not be able to perform their desired tasks. In other words, when you are composing messages for the international audience, you should focus on usability—that is, creating texts that are easy to read, understand, and use in a range of cultural environments. From a usability perspective, your text should be easy for your readers to read, understand, and use in their own cultural contexts and locations.

Now, let us consider another scenario. You are an American student who just arrived in the UK for a study abroad program. After arriving on campus, you decide to deposit some money in a nearby bank. As you enter the bank, the supervisor asks if you would like to open a "current account." You are unsure what to say because you have no idea what the "current account" means. When the supervisor clarifies, you understand that "current account" means "checking account" in your home country. While completing the application form, you are asked to enter your date of birth, which you write as "11/09/2003" (i.e., November 09, 2003) just as you would in the US. However, the supervisor misinterprets it as "September 11, 2003" due to the UK's date format being "day/month/year" (09 November 2003) instead of the "month/day/year" format used in the US. Consequently, you are requested to complete a new application form with your date of birth formatted according to the UK standard.

Both this example of the bank account application and the prior example of the printer instructions involve usability. Both also reveal that rhetoric—or the process of creating messages that address expectations and needs of your intended audiences—can be central to usability in intercultural, cross-cultural, and/or international situations even when all parties are native speakers of the same language (e.g., a US student and a UK bank supervisor). In the process of composing texts (print, digital, multimedia, visual) for audiences from other cultures, it is important to recognize how rhetoric plays a crucial role in addressing their needs and expectations.

Both examples also reveal how usability is a matter of understanding the rhetorical expectations of multicultural audiences and creating texts that meet those expectations. For instance, among the key characteristics of professional communication in the US cultural context are directness (the quality of being clear, straightforward, and to the point), brevity (the quality of being brief or short in duration or amount), concision (the quality of being concise by avoiding unnecessary words, redundancies, or wordiness), and precision (the quality of being exact and accurate). But people in Japan often prefer a much more "roundabout approach, in which ideas are conveyed through implication rather than stated explicitly" (Ingre and Basil 7). Essentially, understanding the rhetorical conventions of writing in the target culture—or the culture with which you are sharing information—is essential for composing effective technical documents.

In sum, what works in your cultural context might not work in another. For this reason, it is important to be culturally aware of your audience's rhetorical expectations and needs when creating technical documents for an international audience. By considering these factors, you can produce usable, meaningful, and credible texts that meet the needs of your target audience in the international context.

Understanding Cultural Differences

Differences in cultural rhetorical expectations can result in misunderstanding (a mistaken or incorrect understanding), misperception (a false or inaccurate perception), and misinterpretation (making a wrong interpretation) when sharing information across cultures. From the perspective of international cultural contexts, these aspects can lead to workplace conflicts, poor interpersonal and intergroup relationships, and inefficiency. Therefore, composing texts for multicultural audiences requires a deeper understanding of cultural diversity—the differences in audiences' behaviors, norms, values, and belief systems.

Learning to understand other cultures is valuable and enriching, and can provide new avenues to succeed in today's globalized world. To gain a deeper understanding of audiences in international contexts, you can start by answering four core questions:

- Who are my readers/for what cultures am I writing?
- What do I want my readers to do after they read my document?
- In what context do my readers use my document?
- What type of document am I composing?

If you have clear answers to such questions, you are in a better position to compose usable, meaningful, and credible documents for multicultural audiences.

While writing for other cultures, you should also be careful about the writing conventions of the target culture—the culture for whom you are writing. For example, in Eastern Asia, it's common to start with general descriptions followed by opinions and statements, whereas in the Western world, the main idea is typically stated upfront, followed by supporting details (Li et al. 337). Similarly, when communicating with people in England, it is advisable to use British terms such as "lift," "boot," and "biscuits" instead of American terms such as "elevator," "trunk," and "cookies."

Thus, the more you learn about your audience and the context of use, the better you will be able to generate usable, meaningful, and credible documents. In short, when you write for multicultural audiences, you need to employ more effective approaches for localized usability.

Creating Texts for Localized Usability

Localization is the process of adapting content to support users' requirements in a specific target locale. In simple terms, localization means making your content usable, meaningful, and credible for your intended audience in a specific target culture. Broadly speaking, localization involves creating materials such as print, digital, multimedia, and visual texts to meet the rhetorical expectations and needs of audiences in different cultures. By addressing these expectations and needs, localization makes materials easier for individuals from those cultures to understand and use. In this way, localization is a central approach to usability for intercultural, cross-cultural, and international communication.

Writing for localized usability means avoiding assumptions based on your language, location, and culture. To succeed in writing for localized usability, you need to focus on:

- **Sociocultural factors:** the forces that influence individual's thoughts, feelings, and behaviors within cultures and society
- **Culturally localized user experiences:** users' practical experiences and skills gained through observations or interactions with other people in the local community
- **Local contexts:** the surrounding environmental factors that influence how a product, service, or user interface is perceived and used by people in a particular region or locality

Culture has a major impact on the usability of documents. For instance, users from different cultures may not use a document if it does not align with their behaviors, norms, values, and belief systems. Because culture strongly influences how users perceive your document, understanding your audience's culture and the context of use is vital to the localized usability of your document.

Regardless of document types, producing an effective document typically requires completing the four basic tasks:

- **Deliver ease-of-use information:** because different individuals in different situations have different information needs
- **Use persuasive reasoning or logic:** because readers may disagree with the meaning of the information provided and the action to be taken
- **Weigh the ethical issues:** because unethical communication lacks credibility and promote unethical behaviors
- **Practice good teamwork:** because many organizations produce documents by a team of colleagues with varying skills and experiences

By tailoring your content to the preferences of your intended audiences, you can reduce the risk of misunderstanding information you share with them. The more you understand your audience's rhetorical expectations and cultural needs, the more likely you are to communicate information that is understood and used as intended. In essence, the better you understand the text localization process, the more effectively you can compose usable, meaningful, and credible documents for multicultural audiences in the globalized world.

Document Localization Process

In the globally interconnected world, you are expected to produce documents in ways that audiences from other cultures can understand and use within their local settings or environments. From an international perspective, a usable text is one that supports knowledge, experiences, and skills of users from related cultures. For this, you need to know how you can compose the localized document that can address rhetorical expectations and needs of particular users from particular cultures.

Adopting the document localization process can help you address these factors and produce effective documents for multicultural audiences. As shown in Figure 17.1, you can begin the documentation process with a documentation plan, which entails a preliminary study of the document type and purpose, audience, and context of use. As the documentation project progresses, the next crucial phase involves delving into cultural factors. During this phase, it becomes essential to carefully examine these cultural elements to tailor the materials appropriately, aligning them with the cultural expectations associated with document usability (St.Amant 154). In the third phase, you start drafting and reviewing the document with cultural considerations and contextual awareness of the user's situation.

Figure 17.1: Document Localization Process

At this phase, you thoroughly review the text to ensure its functionality, accuracy, accessibility, and writing quality. While document functionality is concerned with how the document conveys information to its users, accuracy involves correctly presenting information without ambiguity or errors. Likewise, accessible documents ensure that all users, including those with disabilities, receive the same information. Good technical writing avoids unnecessary wordiness, jargon (overtly technical language), and reads smoothly.

As you perform the tasks from 1 to 3, you should also focus on the cultural expectations of rhetoric (presenting information in a credible way), document type (memos, letters, reports, proposals, instructions, reviews, newsletters, presentations, etc.) and the context of use.

Additionally, you need to be aware of other factors that influence the perceptions of the user of your document, including:

- **Content management:** organizing the content based on importance or priority
- **Typography:** the technique of arranging fonts, font sizes, line lengths, line spacing or leading, letter spacing or tracking, and other elements to help readers find the information they need quickly and easily
- **Abbreviations and acronyms:** words formed from initial letters of a phrase and pronounced as a word like NASA
- **Slang, jargon, idioms, humor, and clichés**

Furthermore, you need to carefully implement various rhetorical strategies during this third phase, including design elements (such as page layout and typography) and the principles of design. Some powerful and durable design principles include:

- **Proximity:** putting two related items (a visual that accompanies a text) close to each other so that the reader will interpret them as related to each other
- **Alignment:** placing texts or visuals so their edges line up along the same rows or columns, or their bodies along a common center
- **Contrast:** arranging elements for emphasis and difference (light vs dark, sans serif vs serif)
- **Repetition:** reusing of the same or similar elements throughout the document (e.g., icons, numbers, lines). (Williams 85-94)

You may also consider adding localized components to support regional differences and technical requirements in the target culture for which you are preparing the document.

In your next phase, you can run a usability test to ensure that your document works as intended for your target audience. Usability testing is the process that employs representative participants from the target culture to evaluate the degree to which your document meets their needs and expectations. This process helps you quickly identify any problems with your document and decide how to fix them for localized usability.

To collect users' insights, you can employ various user research methods, such as focus groups, individual interviews, surveys, and ethnography. Ethnographic methods involve gathering information about your document from users in a real-world setting rather than in a usability lab. There are many online resources listed in the Appendix that you can use to learn how to conduct a usability test.

After the test, you come to your final stage. Here, you analyze user feedback to improve your document by revising, reorganizing, and restructuring the contents rhetorically from a user's perspective. In addition to this, you may also need to incorporate other essential elements for publishing and completing the project. This could involve effectively integrating design elements and adhering to design principles that enhance localized usability. To achieve this, you need to consider some guidelines and best practices discussed below.

Guidelines and Best Practices

The way a document is organized may vary from culture to culture. Unlike Canadians, for example, who like recommendations at the beginning of a report, Germans prefer the background first and recommendation later (Varner and Beamer 73). Similarly, unlike German and Japanese email messages, Arabic email messages have a high level of appreciation and praise

(Danielewicz-Betz 30). So, when writing for an international audience, it is important to be culturally aware and sensitive. This entails recognizing and respecting the norms, values, and belief systems of your readers, which can greatly enhance the impact and reception of your message.

Language Expectations

The use of language in your document should be in its most accepted form, adhering to the writing conventions of the target culture where the document is used. English has many standard varieties around the world, such as North American English, British English, Australian English, and Indian English. So, while composing text for an audience from another culture, you are more likely to succeed using common English words, short sentences, and short paragraphs than using high-sounding phrases, long sentences, and long paragraphs.

When writing for international audiences, consider the following:

- **Use plain language** and avoid slang, jargon, idioms, clichés, phrasal verbs, humor (especially, puns), and metaphors (a comparison between two things that are otherwise unrelated).
- **Avoid acronyms.** An international audience may not know, for instance, what "NASA," "ASAP," "FYI," and "SSN" mean unless you define them.
- **Don't use the same word to mean different things** (e.g., lie, wind, bark).
- **Avoid using special characters** such as "/\:;,*? <>" in the file name.
- **Pay attention to punctuation,** such as quotation marks (" ") in English, low quotes („ ") in German, and guillemets (<< >>) in French.
- **Use "international" English** as used by the target audiences in their cultural context. For instance, different spellings in the words such as "localise," "colour," "centre," "licence," and "traveller" are common in the countries like Nepal, UK, and India.
- **Limit pronouns** as they can cause confusion. For instance, the French word "il" could mean "he" or "it," so your French reader may be unclear about your subject.
- **Use appropriate salutations.** For instance, first names are seldom used by those doing business in Nepal or Germany. Unlike in the US cultural context, salutations such as "Dear sir" are more common than "Dear professor" or "Dear Dr. [last name]" in the Nepali cultural context when writing a letter or email to a male professor.

DOCUMENT FORMATTING EXPECTATIONS

As you prepare documents for your international audience, you need to make certain decisions carefully, including how to format the document. Indeed, words are important to any writer, but in technical writing, how the words look on the page is nearly as important as what the words say. If you want your readers to stay focused on your document, you need to make it visually appealing as well. In formatting a document for multicultural audiences, consider how you can:

- Allow for sufficient text expansion based on the requirements and preferences of the target audience. Text expansion occurs when the target language (the language your audience translates into) takes up more space than the source language.
- Consider where you should put your title, headings, and subheadings in your document.
- Leave appropriate margins in your document.
- Consider your audience's preferences when using columns, paragraphs, bullets, and numbers.
- Keep your document's formatting consistent throughout to avoid confusing your target audience.
- Implement the principles of design and design elements (such as font, font size, lower or upper case, line length, line spacing, and color scheme) that are acceptable in the target culture with which you share the document.

EXPECTATIONS FOR VISUALS AND COLORS

Visuals and colors are often used in technical documents in order to enhance meaning (Cohn 25-27). These features also help capture your audience's attention. But you need to use such features carefully, for they might have different meanings in different cultures. For instance, the meanings of colors can vary depending on the culture of the target audience:

- While white in Western culture symbolizes purity, it is associated with death and mourning in many southeast Asian cultures.
- Red commonly indicates warning or danger in North America, Europe, and Japan but it symbolizes good luck and happiness in China.
- In most part of North America, yellow represents happiness and warmth but Latin America sees yellow as a sign of death, sorrow, and mourning. In Germany, yellow is the color of envy. (Dimitrieska and Efremova 80-83)

Color is an important aspect in every culture. So, understanding the meanings of colors in other cultures is crucial when adding color to your document.

Supplementing written texts with visuals such as graphics and photographs can also help achieve conciseness, clarity, and appeal. However, incorporating visuals into a text for another culture requires a better understanding of the implications of those visuals in that particular culture. For instance, Chinese users tend to favor diverting, cartoon-like visuals much more than Western users (Li et al. 150).

While integrating colors and visual elements into your document for multicultural audiences, consider the following:

- Define and explain color symbology and use colors appropriately for the target audiences.
- Use visuals and symbols that are acceptable and meaningful to the target culture.
- When using human figures, consider using simple, abstract figures devoid of recognizable human face and hairstyle. Omit any indication of skin color and use unshaded line drawings of people.
- Avoid using hand gestures unless you are familiar with their specific meanings and cultural significance in the target culture.

Certainly, you may not be able to account for all the different cultural contexts where colors, visuals, and gestures convey different meanings. But conducting user research and usability testing can help you better understand what colors and visuals your audience prefers and why.

Formatting Numbers, Dates, and Currencies

Consider that different cultures use distinct formats for numbers, dates, and currencies. For example, phone numbers and time representations (such as 12-hour vs. 24-hour) vary around the world. To make your technical documents usable, meaningful, and credible in other cultures, you need to consider the following:

- **Date format:** Different countries use different date formats. For example:

U.S.	Ghana	Portugal	Luxembourg
5/29/2023	29/5/2023	23.05.29	29.05.2023
5-29-2023	29-5-2023	2023/05/29	
5/29/23		29-05-2023	

Units of measurement: Be mindful of the type of measurement units being used in other cultures. For instance, while "kilometer" is commonly used in countries like Nepal, India, Pakistan, and China, "mile" is the preferred unit in the United States, Myanmar, Liberia, and Puerto Rico.

- **Currency:** Pay attention to how currency values are formatted and written. For example:

U.S.	Nepal	India
$430.50	रू ४३०.५०	₹४३०.५०
US$430.50	४३० रुपैयाँ ५० पैसा	४३० रुपये ५० पैसे

- **Large numbers:** Commas and/or periods are used in different places in different countries. For example,

   1,234,567.89 (US)   1.234.567,89 or 1234 567,89 (France)
   12,34,567.89 (Nepal)

- **Paper size:** Consider the paper size your target audience uses, especially for electronic documents. For example, in Nepal, A4 paper (210 x 297 mm, or 8.27 x 11.7 inches) is used in place of American letter-size (8.5 x 11 inches).

Thus, different cultures may perceive your composition of information in a variety of unexpected ways, so carefully consider the organization of contents, language expectations, colors, visuals, and measurement units, just to mention a few.

Researching Other Cultural Expectations

Learning about, and paying close attention to, your audience and the context of use are vital to composing a text for localized usability. Of course, you might face some challenges for preparing technical documents for multicultural audiences. For instance, you might not know how to conduct user research, where to go to review related materials, and what technical tools can be employed to get user feedback on your document. However, various approaches can be employed to explore diverse cultural expectations and incorporate them into the document-writing process to enhance localized usability. Here are some suggestions that you might find helpful to implement the ideas discussed in the chapter.

Review Materials Created by and for Other Cultures

One of the best ways to learn about user expectations in other cultures is to review materials created by and for those cultures. For instance, if you are preparing a document in which you need to include information about foreign exchange rates in Nepal, you might review the website of Nepal Rastra Bank (https://www.nrb.org.np). Similarly, if you are writing an instruction manual for Portuguese (Brazilian) audiences, you might consult the Microsoft Portuguese (Brazilian) Style Guide (https://tinyurl.com/5n8t8z7h). In short, reviewing materials generated by and for other cultures can help you gain a general, initial understanding of the expectations of those cultures.

Do Interviews with Users from the Target Culture

Interviews are a frequently used method to gain insights into your target audience. To gather valuable feedback from intended users, you can conduct interviews either in person or remotely, using phone or popular video conferencing platforms such as Zoom, GoToMeeting, and Skype. Preparing for an interview involves several key steps, including selecting the appropriate interview type, crafting well-considered questions, creating necessary materials, and recruiting potential participants. You need to design interview questions carefully and focus on how to interact with participants to get high-quality data. The process involves designing questions to recruit test participants, such as:

- How often do you use technical documents?
- What types of documents do you use?
- What makes you use them? Is there any reason for it?
- Would you prefer print or online documents?
- How does using print [or online] documents benefit you the most?

The resulting answers help you determine if the test participant is the appropriate person to collect feedback on your document. Gathering reliable data requires recruiting representative participants; otherwise, your results will not help you prepare the document effectively for your intended audience.

Conduct User Tests of Draft Texts with Users from the Target Culture

Conducting user tests is one of the best ways to gain insights into what users need and expect from your document in a localized context. By performing user tests, you can identify areas where your document can be improved for localized usability, such as by using more appropriate language, cultural references, and design elements. To better understand how to implement user tests, you can consult some of the online sources listed in the Appendix. If you are conducting the test online, you can share the document by Skype, GoToMeeting, or Zoom's "share screen" feature, or have them review it in advance. The process involves designing certain questions such as,

- What is your first impression of this document?
- Can you describe what this document is about?
- What convinces you to use this document?
- Do you find the document easy or difficult to use? Why or why not?
- Can you explain what [X] means to you in the document?
- What parts of the document did you like the most? Why?
- What parts of the document did you like the least? Why?

The answers obtained can guide you in revising and adapting the document to align with the preferences and usability expectations of audiences from different cultural backgrounds.

Concluding Thoughts

Composing texts for localized usability requires you to better understand how cultural factors influence your readers' perception of the messages or texts you share with them. Cultures have different expectations and needs associated with the structure of the content, layout, style, mechanics, lexicon, spelling, and other factors. Successfully localized documents should appear as if they originated in the local culture. Miscommunication between cultures can occur due to ineffective localized usability implementation in a document.

Composing technical documents for localized usability, however, can be challenging—especially if you are uncertain about how to obtain information about your audience's behaviors, norms, values, and belief systems. The culture of your target audience has a significant impact on how users perceive and accept your message. Adopting the document localization process as a tool and researching audiences' needs and expectations in the target culture can help produce usable, meaningful, and credible documents. In short, with user research and usability testing in international contexts, you can better understand how cultural differences affect the audience's expectations of your text and compose better texts for better experiences.

Works Cited

Ball, Donald A. and Wendell H. McCulloch. *International Business: The Challenge of Global Competition*, 7th ed., Mcgraw-Hill, 1999.

Cohn, Janae. "Understanding Visual Rhetoric." *Writing Spaces: Readings on Writing*, vol. 3, edited by Dana Driscoll et al., WrtingSpaces.org, Parlor Press, and The WAC Clearinghouse, 2020, pp. 18-39. https://wac.colostate.edu/docs/books/writingspaces3/cohn.pdf.

Danielewicz-Betz, Anna. "(Mis) Use of Email in Student-Faculty Interaction: Implications for University Instruction in Germany, Saudi Arabia, and Japan." *JALT CALL Journal*, vol. 9, no. 1, 2013, pp. 23-57.

Dimitrieska, Savica, and Tanja Efremova. "Colors in the International Marketing." *Entrepreneurship*, vol. 9, no. 1, 2021, pp. 78-86. https://doi.org/10.37708/ep.swu.v9i1.7.

Ingre, David, and Robert Basil. *Engineering Communication: A Practical Guide to Workplace Communications for Engineers*. Cengage, 2016.

Lebrón, Antonio. "What is Culture?" *Merit Research Journal of Education and Review*, vol. 1, no. 6, 2013, pp. 126-132.

Li, Qian, et al. "Getting the picture: A Cross-cultural Comparison of Chinese and Western Users' Preferences for Image Types in Manuals for Household Appliances." *Journal of Technical Writing and Communication*, vol. 51, no. 2, 2021, pp. 137-158, https://doi.org/10.1177/0047281619898140.

Li, Qian, et al. "Inductively Versus Deductively Structured Product Descriptions: Effects on Chinese and Western Readers." *Journal of Business and Technical Communication*, vol. 34, no. 4, 2020, pp. 335-363, https://doi.org/10.1177/1050651920932192.

Redish, Janice C. Ginny. "What is Information Design?" *Technical Communication*, vol. 47, no. 2, 2000, pp. 163-166.

St.Amant, Kirk. "Writing in Global Contexts: Composing Usable Texts for Audiences from Different Cultures. "*Writing Spaces: Readings on Writing*, vol. 3, edited by Dana Driscoll et al., WrtingSpaces.org, Parlor Press, and The WAC Clearinghouse, 2020, pp. 147-61. https://wac.colostate.edu/docs/books/writingspaces3/stamant.pdf.

Williams, Robin. *The Non-designer's Design Book: Design and Typographic Principles for the Visual Novice*, 4th ed., Pearson, 2015.

Varner, Irish and Linda Beamer. *Intercultural Communication in the Global Workplace*, 5th ed., McGraw-Hill, 2011.

Teacher Resources

Overview and Teaching Strategies

Today, students work in a creative, robust technological environment with a variety of media outlets. By using these outlets, they are expected to compose texts for a diverse range of audiences. Communicating effectively with an audience from another culture requires understanding how cultural expectations and needs can affect the way audiences receive and interpret texts. For instance, Western user documentation can differ significantly from Eastern Asian user documentation in terms of content organization, layout, expectations for colors and visuals, and formatting of dates or numbers. To address these challenges, students can adopt more effective composing strategies to create texts that multicultural audiences can find usable, meaningful, and credible.

As discussed in the chapter, students can adopt the document localization process as a tool to compose documents for localized usability. Through a number of examples, the essay explores the challenges and opportunities of understanding and adapting to cultural differences while composing technical documents for multicultural audiences in an international context.

When writing for audiences across cultures, students need to know that different cultures have different needs and preferences. Document types that are popular in one culture may not be popular in another. For example, a message for a Spanish audience might be different from a message for a Nepali audience. This is due to the different cultural norms, values, and belief systems that people have. For these reasons, students should view technical writing tasks as more than just sharing information with an audience; they should also be seen as an opportunity to understand cultural relevance and the rhetorical nature of localization.

This essay is ideally suited for teaching students how to compose technical documents for audiences from other cultures. It provides a comprehensive list of guidelines and practices for writing effective technical documents in the international context, as well as ideas for conducting research on other cultural expectations. Students can use this chapter and adopt the documentation localization process to gain a deeper understanding of how cultural factors impact their writing practices. Instructors can use this entry to convey these ideas and have students practice composing usable, meaningful, and credible technical documents for international audiences.

Discussion Questions

To help students explore the ideas discussed in this chapter, consider having them address—as individuals, in small groups, or as an overall class—the following questions:

1. As demonstrated in the chapter, misinterpretations of text can arise in diverse cultural contexts. Have you ever encountered such situations due to cultural differences?

2. How can the use of color, visuals, layout, typography, organization, structure, and style be optimized in a technical document to effectively achieve informative and persuasive goals across diverse cultural and contextual settings?
3. How can information be managed for localized usability? Can you provide some examples of the best practices of audience analysis to create usable, meaningful, and credible documents in the international context?
4. This entry discusses the document localization process as a tool to create more effective texts for an audience from another culture. What challenges do you anticipate while implementing the document localization process in a text for an international audience?
5. This essay provides some guidelines and best practices to compose texts with cultural considerations. What additional guidelines and practices can you suggest for composing texts with cultural considerations? Can you prepare your own list?
6. Briefly explain the significance of the following while composing texts for multicultural audiences:
    o Cultural expectations
    o User research
    o Document localization
    o Usability testing
    o Localized usability
7. Discuss what strategies would you consider for composing usable, meaningful, and credible technical documents to be used by members from other cultures? Can you provide any examples?

Appendix

- Usability Testing Guidelines: https://www.plainlanguage.gov/guidelines/test/usability-testing/
- Usability Evaluation Basics: https://www.usability.gov/what-and-why/usability-evaluation.html
- Usability Resources: https://infodesign.com.au/usabilityresources.html
- Usability Testing: https://www.nngroup.com/articles/usability-testing-101/
- Remote Testing: https://www.usability.gov/how-to-and-tools/methods/remote-testing.html
- Usability Testing Basics: https://studentlife.mit.edu/sites/default/files/Documents/Usability_Testing_Workshop_March2020.pdf
- How to Test the Usability of Documents: https://www.uxmatters.com/mt/archives/2020/05/how-to-test-the-usability-of-documents.php

Contributors

Keshab Raj Acharya is affiliated with the University at Buffalo—the State University of New York—where he teaches courses related to technical communication in the Department of Engineering Education. His research spans a range of technical and professional communication topics, including usability studies, user localization, information design, cross-cultural technical communication, international technical communication, and social justice. His publications have appeared in the journals *Technical Communication*, *Journal of Technical Writing and Communication*, *Technical Communication and Social Justice*, *Communication Design Quarterly*, and *Present Tense*, among others.

Audrey G. Bennett is University of Michigan inaugural University Diversity and Social Transformation Professor. She is also a former Andrew W. Mellon Distinguished Scholar of the University of Pretoria, South Africa. She studies the design of transformative images that, through interactive aesthetics, can permeate cultural boundaries and impact how we think and behave towards good social change. Awarded with the 2022 AIGA Steve Heller Prize for Cultural Commentary, her research publications include "How Design Education Can Use Generative Play to Innovate for Social Change," *International Journal of Design*; *Engendering Interaction with Images*; "The Rise of Research in Graphic Design," *Design Studies: Theory and Research in Design*; "Interactive Aesthetics," *Design Issues*; and "Good Design is Good Social Change," *Visible Language*.

Felicia Chong is a UX researcher with an academic background in rhetoric and technical communication. She has taught classes in business writing, composition, editing, science writing, technical writing, and usability. Her work has appeared in *IEEE: Transactions in Professional Communication*, *Programmatic Perspectives*, *Technical Communication*, and *Technical Communication Quarterly*. She received the 2022 Nell Ann Pickett Award for her co-authored article, "Student Recruitment in Technical and Professional Communication Programs," in *Technical Communication Quarterly*.

Yvonne Cleary is Associate Professor in Technical Communication and Instructional Design at the University of Limerick, Ireland, where she serves as Head (Chair) of English, Irish and Communication. She teaches courses in technical communication theory and practice, writing, and e-learning. Her research interests include professional issues in technical communication, virtual teams, and international technical communication. Her recent book, *The Profession and Practice of Technical Communication*, is published with Routledge.

K. Alex Ilyasova is Associate Professor in the Technical Communication and Information Design (TCID) program and an Associate Dean of the College of Letters, Arts, and Sciences at the University of Colorado, Colorado Springs. She has served as a writing program administrator for ten years, and her research interests include program administration, and empathy and emotional intelligence in TC and UX. Her articles have appeared in *Technical Communication Quarterly,* and *Programmatic Perspectives*, and in a number of edited collections.

Laurence José is Associate Professor of Writing and the Director of the Digital Studies minor at Grand Valley State University where she teaches courses in professional writing, visual rhetoric, and digital studies. Her research interests include cross-cultural communication, data design, genre theory, semiotic theory, and program administration. Her work has appeared in *The Nordic Journal of English Studies,* in various edited collections, as well as in French linguistics journals, including *SCOLIA, Langages,* and *Le Bulletin de la Société de Linguistique de Paris.*

Clint Lanier is Assistant Professor of English at New Mexico State University where he teaches rhetoric and professional communication. His research has been featured in numerous technical communication journals and books.

Candice Lanius is a Principal User Experience Researcher/Human Factors Engineer at Northrop Grumman. Her research interests are in UX, user centered design, and rhetoric of data. This publication stems from her previous faculty role in the Communication Arts Department at the University of Alabama in Huntsville.

Birgitta Meex is Assistant Professor of German and Applied Linguistics at KU Leuven University, where she teaches German grammar, business German and Technical and Professional Communication (TPC). Her research focuses on writing and reading processes in technical communication, writing for localization, the use of language in UX design, technical communication pedagogy, and discourse-analytical aspects of technical communication. She has published essays in *Lebende Sprachen*, *The Journal of Internationalization and Localization, Journal of Technical Education*, and several edited collections. She is a founding member and the delegate of tekom Belgium and a tekom Certified Trainer in Technical Communication.

Cathryn Molloy is Professor of Writing Studies in the University of Delaware's English department. She is the author of *Rhetorical Ethos in Health and Medicine: Patient Credibility, Stigma, and Misdiagnosis* and is co-editor of the *Rhetoric of Health and Medicine* journal.

Therese I. Pennell is Assistant Professor and Director of the Writing Center at Prairie View A&M University in Prairie View Texas. Dr. Pennell's works focus on intercultural communication. She has published individually and collaboratively on topics such as online pedagogy, professional mentorship, visual design, social justice, and intercultural technical communication. Some of her works have been published in *Programmatic Perspectives*, *Texas Women in Higher Education*, and *Journal of Technical Writing and Communication*.

Daniel P. Richards is Associate Professor of English at Old Dominion University, where he serves as Associate Chair and director of the graduate and undergraduate programs in technical and professional writing. His essays have appeared in *Technical Communication Quarterly*, *Journal of Business and Technical Communication*, *Communication Design Quarterly*, and *Contemporary Pragmatism*.

Tammy Rice-Bailey is Associate Professor of Technical Communication (TC) and User Experience (UX) at the Milwaukee School of Engineering. Prior to her academic career, she managed corporate documentation and training for global software, financial, and retail corporations. She has been published in *Technical Communication*, *Technical Communication Quarterly*, *Communication Design Quarterly*, and *Technical Writing and Communication*. She is also the co-author of the forthcoming book, *Interpersonal Skills for Group Collaboration*.

Emma J. Rose is Associate Professor in Interdisciplinary Arts & Sciences at the University of Washington Tacoma and has a Ph.D. in Human Centered Design & Engineering. Her research is motivated by a commitment to social justice and a belief that the way technologies are designed, and who is doing the designing, helps shape our world. Her research interests include the practice and pedagogy of user experience, participatory research, and community based UX design. Her work has appeared in industry and scholarly publications including *UXPA* magazine, *Communication Design Quarterly*, *International Journal of Human-Computer Interaction*, and *Technical Communication Quarterly* among others.

Joanna Schreiber is Associate Professor of Professional and Technical Communication at Georgia Southern University. Her research interests include project management, trends in professional and technical editing, workplace studies, and UX methods. Her work has been published in *Technical Communication*, *Technical Communication Quarterly*, *Programmatic Perspectives*, and *Journal of Technical Writing and Communication*.

Darina M. Slattery is Associate Professor at the University of Limerick (UL), where she specializes in e-learning and instructional design. Since the start of 2021, Darina has been serving in leadership roles for the UL Learning Management System (LMS) Implementation Project. Darina won the UL Teaching Award for Excellence in the Provision of Pedagogical Support in 2019 and the STC's Frank R. Smith Award for Outstanding Journal Article (with A. Cudmore) in 2020. She served as President of the IEEE Professional Communication Society (ProComm) from 2021-2022 and in 2023, she was awarded the Emily K. Schlesinger Award for Outstanding Service to the IEEE Professional Communication Society.

Ryan Weber is Associate Professor of English and the Director of Business and Technical Writing at The University of Alabama in Huntsville. His work has appeared in *Technical Communication Quarterly*, *Communication Design Quarterly*, *Journal of Technical Writing and Communication*, and other journals. He also hosts the podcast *10-Minute Tech Comm*.

Candice A. Welhausen is Associate Professor of Technical and Professional Communication and Associate Chair of the English Department at Auburn University, where she teaches classes in technical and professional communication, visual communication and information design, and science and grant writing. Before becoming an academic, she was a technical writer/editor at the University of New Mexico Health Sciences Center. Her workplace experience directly informs her teaching and research as well as her overall approach to the theory and practice of technical communication.

Quan Zhou is Professor in the Department of Technical Communication and Interaction Design at Metropolitan State University, where he directs the Design of User Experience programs. His essays have appeared in *Technical Communication* and *Communication Design Quarterly*, and he is the editor-in-chief of the *Journal of Rhetoric, Professional Communication, and Globalization*.

www.ingramcontent.com/pod-product-compliance
Lightning Source LLC
Chambersburg PA
CBHW032359230426
43672CB00007B/752